T0238726

Communications
in Computer and Information Science 999

Commenced Publication in 2007
Founding and Former Series Editors:
Phoebe Chen, Alfredo Cuzzocrea, Xiaoyong Du, Orhun Kara, Ting Liu,
Krishna M. Sivalingam, Dominik Ślęzak, and Xiaokang Yang

Editorial Board

Simone Diniz Junqueira Barbosa
 Pontifical Catholic University of Rio de Janeiro (PUC-Rio),
 Rio de Janeiro, Brazil
Joaquim Filipe
 Polytechnic Institute of Setúbal, Setúbal, Portugal
Ashish Ghosh
 Indian Statistical Institute, Kolkata, India
Igor Kotenko
 St. Petersburg Institute for Informatics and Automation of the Russian
 Academy of Sciences, St. Petersburg, Russia
Takashi Washio
 Osaka University, Osaka, Japan
Junsong Yuan
 University at Buffalo, The State University of New York, Buffalo, USA
Lizhu Zhou
 Tsinghua University, Beijing, China

More information about this series at http://www.springer.com/series/7899

Fernando Koch · Atsushi Yoshikawa ·
Shihan Wang · Takao Terano (Eds.)

Evolutionary Computing and Artificial Intelligence

Essays Dedicated to Takao Terano
on the Occasion of His Retirement

 Springer

Editors
Fernando Koch 🄭
University of Melbourne
Melbourne, VIC, Australia

Shihan Wang 🄭
University of Amsterdam
Amsterdam, Noord-Holland
The Netherlands

Atsushi Yoshikawa 🄭
Tokyo Institute of Technology
Tokyo, Japan

Takao Terano 🄭
Chiba University of Commerce
Tokyo, Japan

ISSN 1865-0929   ISSN 1865-0937 (electronic)
Communications in Computer and Information Science
ISBN 978-981-13-6935-3   ISBN 978-981-13-6936-0   (eBook)
https://doi.org/10.1007/978-981-13-6936-0

Library of Congress Control Number: 2019935138

© Springer Nature Singapore Pte Ltd. 2019
This work is subject to copyright. All rights are reserved by the Publisher, whether the whole or part of the material is concerned, specifically the rights of translation, reprinting, reuse of illustrations, recitation, broadcasting, reproduction on microfilms or in any other physical way, and transmission or information storage and retrieval, electronic adaptation, computer software, or by similar or dissimilar methodology now known or hereafter developed.
The use of general descriptive names, registered names, trademarks, service marks, etc. in this publication does not imply, even in the absence of a specific statement, that such names are exempt from the relevant protective laws and regulations and therefore free for general use.
The publisher, the authors and the editors are safe to assume that the advice and information in this book are believed to be true and accurate at the date of publication. Neither the publisher nor the authors or the editors give a warranty, expressed or implied, with respect to the material contained herein or for any errors or omissions that may have been made. The publisher remains neutral with regard to jurisdictional claims in published maps and institutional affiliations.

Cover illustration: The image on the front cover is the work of © Management Service Center Co., Ltd. All rights reserved. Used with permission granted on February 19, 2019.
Photograph on p. v: The photograph of the honoree is from Prof. Takao Terano's personal archive. Used with permission.

This Springer imprint is published by the registered company Springer Nature Singapore Pte Ltd.
The registered company address is: 152 Beach Road, #21-01/04 Gateway East, Singapore 189721, Singapore

Prof. Dr. Takao Terano

Preface

This *Festschrift*, "Letters of the Special Discussion on Evolutionary Computing and Artificial Intelligence," is published on the occasion of the retirement of Prof. Dr. Takao Terano and contains the results and invited papers from the General Conference on Emerging Arts of Research on Management and Administration (GEAR)[1], held on April 14, 2018, at the Tokyo Institute of Technology, Suzukakedai, Kanagawa, Japan.

Artificial intelligence and evolutionary computing are technologies that have been in vogue for a long time, going through their winters and hypes. These technologies are in the forefront of new tools and applications, creating actionable insights, real-time awareness, and predictive analysis of numerous topics for sustainable development and humanitarian action. AI is deemed to be a key technology to support new tools that will address challenges of the United Nation's Sustainable Development Goals (SDGs), as presented in the AI for Good Global Summit 2018[2]. Compelling examples illustrate the value of this technology for improving early warning systems and inform policy and programmatic response.

In this context, GEAR aimed to to discuss the progress of non-traditional and cutting-edge methodologies and applications in management and administrative issues. The conference focuses on experiment-based management and administrative science from both methodological and applied aspects.

This volume includes extended and revised versions of the selected work, complemented with invited papers from authors with knowledge in the field. The workshop promoted an international discussion forum with Program Committee members from many countries in Asia (Japan, China, South Korea), Europe (Greece, Germany, The Netherlands), Oceania (Australia, New Zealand), and the Americas (USA, Brazil).

GEAR received 23 submissions through the workshop website, from which we selected 15 for presentation; nine extended versions of these contributions are published as chapters of the current volume, along with invited papers and a special contribution from Prof. Terano. All the submissions were single-blinded reviewed by at least three members of the Program Committee.

In the first chapter, "This Is How I Feel About Complex Systems," Prof. Takao Terano describes his 40+ years experience in the field of computer sciences, details how he feels about complex systems, and discusses principles of agent-based modelling, interdisciplinary research, and system creation. This paper is a summary of his last class at Tokyo Institute of Technology.

In the second chapter, "Artificial Intelligence Technology and Social Problem Solving," Kim and Cha present the program of the Korean government for mega-projects to address low employment, population ageing, low birth rate and social safety net problems by utilizing AI and ICBM (IoT, cloud computing, big data, mobile)

---

[1] GEAR 2018: https://sites.google.com/view/gear2018en/home.

[2] AI for Good Global Summit 2018, ITU, UN: https://www.itu.int/en/ITU-T/AI/2018/.

technologies. The paper presents the authors' views on how AI and ICT technologies can be applied to ease or solve social problems by sharing examples of research results from studies of social anxiety, environmental noise, mobility of the disabled, and problems in social safety.

Next, in "A Formal Model of Managerial Decision Making for Business Case Description," Kunigami, Kikuchi, and Terano propose the managerial decision-making description model as a novel formal description model of organizational decision-making. The model introduces a four terminal element representation to describe managerial decisions redefining relationships between their objectives and resources. The paper demonstrates how the model is applicable to business and in facilitating business gaming.

In the fourth chapter, "Evolutionary Computation Meets Multi-Agent Systems for Solving Optimization Problems," de Carvalho and Sichman discuss the synergy between evolutionary computation and multi-agent systems used together to enhance the process of solving optimization problems. The papers present an overview of this combined approach, considering both mono- and multi-objective approaches, and stress the importance of hyper-heuristic approaches.

In the fifth chapter, " Practice Report on Active Learning Using Business Games for Teaching Training Course Students," Uchida and Yuasa describe a proposal to make students experience business games using computer agents and consider alternatives for the utilization of technology in education.

Extending the education field, in the sixth chapter, "Beyond Educational Policy Making," Yoshikawa and Takahashi present an innovative method to incorporate agent-based simulation (ABS) into the policy formulation process around education. The claim is that the formulation of educational policy is based on data that can turn out to be difficult to access. The work presents case studies demonstrating that the risk can be reduced by incorporating ABS into the policy formulation process. The paper presents ABS description levels, and discusses risks that both can and cannot be expressed using ABS.

The seventh chapter, "Framework of Evaluating Business Partner Recommendation Beyond Industry Types Toward Virtual Corporation," by Mukai proposes a framework to evaluate a recommendation of unknown partners in an inter-business market by artificial intelligence (AI) and simulation. The framework consists of a sequence of steps to (a) propose a method of recommending unknown business partners, (b) install the recommendation method as AI into a firm agent, and (c) evaluate a recommended business partner by comparing the performance between a recommendation method. The framework provides a decision recommendation service for company managers.

In the eighth chapter, "Analysis of Researchers Using Network Centralities of Co-authorship from the Academic Literature Database," Fujita introduces a quantitative method that satisfies the requirements to evaluate researchers in organizational R&D fields. The results suggest the demand for competencies in both expertise in the research fields and cooperativeness with others in the projects. The work proposes a new quantitative method to evaluate researchers by measuring the network centralities from the academic literature database.

The ninth paper, "Debriefing Framework for Business Games Using Simulation Analysis," Kikuchi et al. outline a framework to support the evaluation of player behavior in technology-based business games. The analysis involves two steps: (a) constructing an agent-based model based on the subject of the business game and categorizing simulation-log, and (b) mapping logs of player behavior onto typed results. The solution supports visualizing payers' positioning along with a range of possible results. This allows for providing useful debriefing information for both players and facilitators.

The tenth paper, "Applications of Evolutionary Computation and Artificial Intelligence in Metallurgical Industry," by An, She, Chen, and Wu, describes how the development of evolutionary computing and artificial intelligence impacts the metallurgical industry by analyzing some good applications in typical metallurgical processes. The work discusses the future development trends and challenges in the applications of computational intelligence-based technologies in metallurgical industries.

Next, in "Evolutionary Computation and Artificial Intelligence for Business Transactions" Gotsias discusses the utilization of evolutionary computing and artificial intelligence in business transactions, focusing on issue of synchronization and coordination of agents' activities in supply-chain environments. After briefly discussing the general coordination model, the hourglass model, we present a mathematical model for achieving coordination inside the firm and show how the agent activities are coordinated in a department, as well as across departments. The coordination model specifies the synchronization conditions by considering message travel times and product/support operational requirements. The conditions for achieving coordination and the relationships between operational and support departments are an original contribution in the economics of the firm. In the final part of the paper, we indicate how the coordination model results can be utilized in a dedicated AI environment for studying economic relations among firm/market participants.

Finally, in "Proposal for Mutual Collaboration Between Simulation and Field Research in Archeology" Sakahira proposes a methodology to combine agent-based simulation and conventional research methods in archeology. The method presented encompasses hypothesis verification from simulation results to become the input of archeological field research. The paper presents an application example in the case of determining whether native Jomon people or Chinese-Korean immigrants played the major role in agricultural culture in the Yayoi period.

Accolade

Prof. Dr. Takao Terano is Professor at Chiba University of Commerce, Professor Emeritus at the Tokyo Institute of Technology, and Professor at the University of Tsukuba. He is also an invited researcher of the National Institute of Advanced Industrial Science and Technology (AIST), Japan. He received a BA degree in Mathematical Engineering in 1976, and an MA degree in Information Engineering in 1978, both from the University of Tokyo, and Doctor of Engineering Degree in 1991 from Tokyo Institute of Technology. His interests include gent-based modelling, knowledge systems, evolutionary computation, and service science. He is a member of the editorial board of major artificial intelligence- and system science-related academic societies in Japan, and is a member of the IEEE and president of the PAAA.

Fig. 1. Prof. Dr. Takao Terano, 2018

With more than 40 years of an intense career, it is safe to say that Prof. Terano is one of the most influential researchers in applied artificial intelligence in Japan. His name is recognized in the field worldwide, as confirmed by the *praise* of colleagues around the world:

Prof. Dr. Yeunbae Kim, Institute for Information and Communication Technology Promotion, South Korea:

I had the opportunity and privilege of working with Prof. Terano in projects involving AI technologies, and I was deeply impressed not only by the impeccable work done by Prof. Terano but also by his warm leadership. Prof. Terano's contributions to the scientific community are immeasurable, and many of them helped define new research directions in the AI and social intelligence fields. Your presence will be dearly missed, but I wish you the best in your future endeavors and I hope you have a fun and fruitful retirement!

Prof. Dr. Masayuki Yamamura, Tokyo Institute of Technology, Japan:

I am very happy to hear about the CCIS Festschrift volume on the occasion of the retirement of Prof. Takao Terano. Prof. Terano has been one of the world's most influential researchers in the area of applied artificial intelligence. His contribution and outreach are widely recognized in the research community. I was a student when he attained his doctoral degree in Professor Kobayashi's Laboratory at the Tokyo Institute of Technology. We have been good colleagues since he became a professor in the same department. Although our research field is slightly different, we were both involved in judging students' thesis defences.

Prof. Jaime Simão Sichman, University of São Paulo, Brazil:

Prof. Terano has been a reference for me and other researchers in the domain of agent-based social simulation. Being one of the founders of the PAAA (Pan-Asian Association for Agent-Based Approach in Social Systems Sciences), he has proposed and has actively organized the AESCS (Agent-Based Approaches in Economic and Social Complex Systems) workshop series since 2001. We met each other at the 4th European Social Simulation Association Conference, held in Toulouse in 2007, and this partnership has evolved and led to our common work in the organization of the 5th World Congress on Social Simulation (WCSS 2014), held in Sao Paulo, Brazil, during November 4–7, 2014, where I had the role of general chair and Prof. Terano was one of the program co-chairs. For me, it was a great pleasure to work with him, and I hope this partnership can happen again in the very near future.

Fig. 2. © Management Service Centre Co., Ltd. All rights reserved

Prof. Dr. Gotsias Apostolos, Department of Business Administration, University of the Aegean, Greece:

Professor Terano has held and will continue to hold for many years to come the beacon of far-reaching light into explored and unexplored scientific areas. He has been a forerunner in applied artificial intelligence and in applied business intelligence environments using Manga. It was a great pleasure and honor for me to explore such new areas through the beacon he holds for others and for me.

Dr. Fernando Koch, co-editor, IBM, The University of Melbourne, Australia:

Prof. Terano is one of the world's influential researchers in the area of applied artificial intelligence. He was an *AI digital influencer* in Japan long before this term became a hit. His contribution and outreach are widely recognized in the research community. I recall my first contact with Prof. Terano was through Samsung Research, who contracted Terano's Lab[3] research for strategic projects in the area of social engagement and digital education. Since then, our collaboration grew fruitfully through multiple workshops and publications. I hope Prof. Terano continues to contribute in mentorship and ideation to the field whilst enjoying a long and pleasurable retirement.

Dr. Atsushi Yoshikawa, co-editor, Tokyo Institute of Technology, EduLab Inc., Japan:

Prof. Terano has been researching in artificial intelligence (AI) since the 1970s when this field started in Japan. Since then he has been a leading figure in the Japanese AI research community. I have been working with Prof. Terano since 2004 to study diverse research themes related to the area. His method focuses on real-world problems. Our laboratory deals with the challenges of adopting AI in areas as diverse as economics, management, education, sports, and so on. This method aims to attract

---

[3] Terano's Lab: http://www.trn.dis.titech.ac.jp/GEAR/index.html.

collaborative research and industry engagement, which has been very successful. Working with Prof. Terano is an exciting journey, full of experiences and new learning.

Dr. Shihan Wang, co-editor, University of Amsterdam, The Netherlands:

Prof. Terano has been well-known for his research in evolutionary computation and artificial intelligence, which also attracted me to join his lab as a doctoral student. Prof. Terano inspires me all the time, and he has patiently guided me to start an interdisciplinary research journey in computational social science. He acts as both the academic and spiritual mentor in my research life.

November 2018

Fernando Koch
Atsushi Yoshikawa
Shihan Wang
Takao Terano

Organization

Organizing Committee

Fernando Koch	The University of Melbourne, Australia
Atsushi Yoshikawa	Tokyo Institute of Technology, Japan
Shihan Wang	Tokyo Institute of Technology, Japan

Steering Committee

Takao Terano	Chiba University of Commerce and Tokyo Institute of Technology, Japan
Yeunbae Kim	Institute for Information and Communication Technology Promotion

Technical Program Committee

Jianqi An	China University of Geosciences, China
Vinicius Renan de Carvalho	University of São Paulo, Brazil
Masanori Fujita	Tokyo Institute of Technology, Japan
Apostolos Gotsias	University of the Aegean, Greece
Takamasa Kikuchi	The University of Tokyo, Japan
Fernando Koch	The University of Melbourne, Australia
Taisay Mukai	Tokyo Institute of Technology, Japan
Fumihiro Sakahira	Kozo Keikaku Engineering Inc., Japan
Jinhua She	Tokyo University of Technology, Japan
Jaime Sichman	University of São Paulo, Brazil
Satoshi Takahashi	Tokyo University of Science, Japan
Hikaru Uchida	Aoyama Gakuin University, Japan
Shihan Wang	Tokyo Institute of Technology, Japan
Atsushi Yoshikawa	Tokyo Institute of Technology, Japan

Conference Committee

Atsushi Yoshikawa	Tokyo Institute of Technology, Japan
Bin Jiang	Hunan University, China
Hikaru Uchida	Aoyama Gakuin University, Japan
Irvan M. Ipan	Tokyo Institute of Technology, Japan
Kazunori Umino	Medical Design Co. Ltd.
Masaaki Kunigami	Tokyo Institute of Technology, Japan
Masakazu Takahashi	Yamaguchi University, Japan
Satoshi Takahashi	Tokyo University of Science, Japan

Sayuri Yoshizawa-Watanabe	Hoshi University, Japan
Setsuya Kurahashi	University of Tsukuba, Japan
Takamasa Kikuchi	Mitsubishi UFJ Trust and Banking and Keio University, Japan
Takashi Yamada	Yamaguchi University, Japan
Tomomi Kobayashi	Kobayashi Institute of Management, Japan
Yuya Murata	National Institute of Advanced Industrial Science and Technology

Acknowledgments

We would like to thank all the volunteers who made the workshops possible by helping to organize and peer review the submissions, and EasyChair for the conference and proceedings management system. We appreciate the help and dedication of the members of the GEAR research communities, who continuously participate in our activities as Technical Program Committee members, submitting contributions, and helping us in putting the pieces together to promote this publication.

We are also grateful to Springer for the continuous support and providing the venue for publishing the printed proceedings after our workshops. This contribution is invaluable in further promoting the research around Artificial Intelligence and Evolutionary Computing.

During the implementation of this project, Dr. Koch was working with the School of Computing and Information Systems, The University of Melbourne (UoM), sponsored by a Global Research Outreach grant from Samsung Advanced Institute of Technology (SAIT), collaboration Project IO170924-04695-01. Dr. Koch is also supported by the CNPq Productivity in Technology and Innovation Award, grant CNPq 307275/2015-9.

Dr. Atsushi Yoshikawa is currently a visiting professor at the Department of Artificial Intelligence with the Tokyo Institute of Technology (TITECH), Japan. He also worked at EduLab inc., Japan. Dr. Yoshikawa is thankful for the assistance provided by TITECH for supporting his participation in the Conference and in this project.

Dr. Shihan Wang is currently a researcher in the Institute of Informatics, University of Amsterdam (UvA), The Netherlands, funded by Netherlands Organisation for Scientific Research (NWO) and Applied Scientific Research (SIA) grant 629.004.013. She is thankful for the support provided by UvA during the implementation of the conference and her participation in this project.

Contents

This Is How I Feel About Complex Systems

Takao Terano[(✉)]

Chiba University of Commerce, Chiba 272-8512, Japan
tterano@computer.org

Abstract. It has been around for nearly 40 years since I joined with the academic world in computer science related areas. Because I prefer to concrete real world topics, I, as a system scientist, usually research and develop challenging cutting-edge applications on complex systems. By complex systems, I do not only mean such areas as complex adaptive systems, chaotic, nor fuzzy systems, but, seemingly complicated systems required to be socially implemented. In this short paper, based on my final lecture at Tokyo Institute of Technology, I would like to present how I feel about complex systems and discuss principles of agent-based modeling, interdisciplinary research, and system creation. The paper concludes some comments on what we must do in the future.

Keywords: Complex systems · Agent-based modeling · System creation

1 Introduction

"Complex systems," as used in the title of this chapter, refers to all social systems. Despite its small scale, I believe that the most challenging social system is household management; it is much easier to debate matters on national or even international scales. "Complex systems" in the broader world are generally complex adaptive systems, but those are slightly narrower in meaning. Chaotic systems also exhibit complex behavior, but describing them is quite simple.

We have an innate desire to simplify any complex matter we face, and through a long history of such self-delusion, we come to believe that we understand the world. Witness the myriad mythologies describing the creation of the sun and the moon; these are merely delusional philosophies designed to provide a sense of understanding. A similar phenomenon occurs in academia. In truth, however, these matters—household issues included—retain their complexity.

My position is that we should actively use the concepts of agent-based modeling and evolutionary computation as new approaches to understanding complex systems. Using these two principles, we can understand complex systems. For further details on how I reached this conclusion, see the description in [1], for example.

© Springer Nature Singapore Pte Ltd. 2019
F. Koch et al. (Eds.): GEAR 2018, CCIS 999, pp. 1–7, 2019.
https://doi.org/10.1007/978-981-13-6936-0_1

2 An Approach to Agent-Based Modeling

Let us discuss the micro-macro links in agent-based modeling. First, an agent is an entity that has an internal state and decision-making and communication functions. Agents can model humans, organizations, or even objects such as molecules. Through the microinteractions of agents, a macroscale order with bottom-up effects emerges. From the standpoint of creating models through which to view the world, we cannot view microscale conditions in detail. Therefore, academia has advanced to the point of observing macroscale phenomena and creating models involving macroscale variables.

Take economics as an example. Currency was created as a result of barter between agents. Currency is continuous and therefore allows for the creation of various equation-based models, which have led to econometric models. I consider this an example of re-creation from zero through agent-based modeling.

Up to this point, natural and social phenomena work in the same way. Go further, however, and they differ. Because agents have internal states and communication functions, they can observe the macroscale order. As a result, top-down influences from the macroscale are transmitted to the micro level, where they alter agent behavior. This is the complex behavior that is generally seen in social phenomena. Once microscale agents begin to observe macroscale phenomena, complex interactions that form micro-macro links arise (Fig. 1).

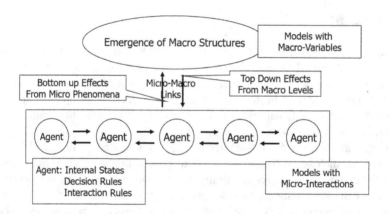

Fig. 1. Framework of agent-based modeling

The desire to program directly in this manner, if possible, led me to agent-based modeling. The programming itself is not all that difficult—the difficulty lies in what comes afterward, in three particular areas.

First, there is a need to skillfully connect theory and reality. For example, game theory experiences problems when applied to the collective behavior seen in real-world social media. Modeling must therefore consider both theory and reality.

As a second issue, this method lies on a disciplinary boundary; therefore, we must be conscious of the differing time scales of the natural sciences, the social sciences, and engineering technology. In the natural sciences, theories are required to be eternally correct within the scope of measurement technology, whereas in the social sciences, we are happy with a theory if it holds for one century. As engineering technicians, however, we expect new technologies to last perhaps a decade; beyond that, they will have been modified to the point of being something else. Such differences in scale become extremely troublesome when creating models.

The third problem is the difficulty of receiving general recognition for the models we create. There are two aspects of this: assessing validity and ensuring correctness. For example, securities exchanges set "limit up" and "limit down" prices to restrain volatility, but the extent to which prices are allowed to fluctuate is arbitrary. There are many other such situations. Although it is possible to somehow mesh models with reality, when considering the design of new structures, determining whether a model works correctly becomes a matter of its persuasiveness, i.e., a matter of its ability to convince. This is extremely difficult to accomplish with agent-based modeling.

3 The Necessity of a Common Language

Still, as mentioned above, it is human nature to desire simplification, even if not to the extent of mythology. This is summarized well in Nobel Laureate Daniel Kahneman's Thinking, Fast and Slow [2]. This book assumes that humans cannot think the unthinkable. Indeed, it repeatedly describes how humans continually avoid thinking. This concept implies that, as we progress in our research, acknowledging such laziness becomes more likely to produce results, and therefore, it is easier to remain in our foxholes. I admit that digging a foxhole is necessary for becoming a specialist in academia, but I also believe that this is probably a bad idea. If we are to escape from our foxholes, we must first identify a common language. If no common language exists, we must create one.

An example of this is, curiously enough, the concept of acupuncture points in Oriental medicine. One might consider these points inherent to the human body, but in fact, they are an invented concept. This is demonstrated by the fact that their different names and positions depend on the tradition. However, whereas the treatment may vary slightly according to the school of thought, using terminology related to acupuncture points allows knowledge to be propagated easily. In other words, the act of naming acupuncture points itself has made it possible to describe treatment methods. Being able to describe treatment methods aids propagation and makes it possible to describe incorrect ways of stimulating a given point (Fig. 2). Therefore, I consider a common language necessary for shared consideration of complex problems. This is what I believe, but it is no easy task in academic areas where progress is fast. Therefore, I consider it necessary to consciously make the effort to create a common language.

Based on the above, cross-domain, or interdisciplinary approaches are vital for social issues in general [3]. As an example, statistical physics has a very powerful method called mean field approximation. This method allows various models to be analyzed and approximate calculations to be performed. To that end, however, the simplifying assumptions are important. In contrast, with agent-based modeling, we can use computers to simulate the behavior of agents and the circumstances of the

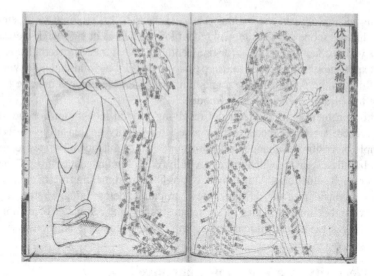

- Acupuncture Points are Invented
- Explanations with the Concepts of them are Essential

Fig. 2. Explanations of acupuncture points in an old Japanese textbook

surrounding world in a bottom-up manner. Computers can understand these descriptions, as can humans with some difficulty, but experiments modeling a world with no particular first principles require substantial parameter tuning.

For example, gravitational models of population dynamics are based on Newton's law of universal gravitation, but their results can be manipulated in any direction using the parameter corresponding to the gravitational constant. When modeling is performed using agents, the movements of people can be observed. That means the two approaches are complementary and must both be appropriately considered when modeling social issues.

4 Creating Systems

I would like to turn the discussion to system creation. Due in part to the recent boom in artificial intelligence, I hear frequent mentions of "advanced systems." Extremely large, elephantine systems are being created, particularly in business fields. In many cases, however, creating an elephant consumes all available resources, which, in turn, leads to businesses relying fully on the system as-is. If the organization includes many intelligent members, they will forbid individuals from interacting with the elephant and instead take parts from it to create trunk systems and tusk systems and ear systems through which business can be conducted. This is not necessarily a bad approach, as long as the business runs smoothly, but the oncoming tsunami of artificial intelligence and big data require those trunks and tusks to be reattached. The result can easily

Fig. 3. A system as an elephant and users as blind people (Photo: Hubei Museum of Modern Art (Wuhan, China))

become a piecemeal creation reminiscent of Frankenstein's monster (Fig. 3). I believe that this is already happening in large companies trying to use big data correctly.

Creating systems has always been an extremely troublesome task, and furthermore, systems tend to become out-of-date as soon as they are completed. Therefore, they need to be broken up and used in parts or tossed aside and re-created. Such re-creation requires time, money, and energy; therefore, executive decision-making comes into play. Taking on a system requires skills related to analysis, design, and imagination. Analysis lies in the field of science. Design and development are in the realm of engineering. Imagination is a matter of sensibility. If all three are not skillfully combined, a good system cannot be produced. We must be aware that in something akin to the "Red Queen hypothesis" from Lewis Carroll's Through the Looking-Glass, one must continue running to avoid becoming increasingly obsolete.

Note that systemization alone is insufficient for solving social problems. In 2016, the Oxford English Dictionary selected "post-truth" as its word of the year, suggesting that we have entered an era that emphasizes impact over truth. For example, today we see a heavy emphasis on the impact of social media. I cannot help but wonder whether, a decade from now, we will be discussing how we were all taken in by social media. Today, we consider fake news and online bullying social issues, but I consider this a positive trend—in the past, only large organizations were capable of manipulating information, but today, individuals have much more power, and post-truth concepts have become highly influential. There are probably various reasons for this increase in individual power, but one significant factor is the advancing capabilities of computer hardware.

5   What We Must Do in the Future

When I consider the state of academia in Japan, I am troubled by many discomforting truths. University researchers and educators are overworked. There is too much competition due to the selectivity and concentration of research budgets. Few large companies have central research laboratories, and corporate researchers have little freedom in choosing their topics.

In fiscal year 2017, the total amount of JSPS KAKENHI grants-in-aid was US$2.1 billion (just under JPN230 billion). This includes amounts dedicated to research on induced pluripotent stem cells and supercomputers. In comparison, the research budget at Amazon is said to be US$22.6 billion. The largest research budget in Japan is that of Toyota at approximately US$9.3 billion. From admittedly rather old data, the total corporate R&D expenditure in Japan is approximately JPN12–13 trillion. Of that, approximately JPN30–40 billion are transferred from companies to universities. There is no way to overcome such vast funding differences, no matter how skilled the researchers are. Thus, Japanese academics are put in an extremely tough spot.

This leads to the question of what we can do about this situation, and, as powerless as we are, I believe that there are two approaches we can take.

The first approach is internationalization, which is more inexpensive now than ever. Even today, research proposals include words to the effect of "they're doing this in the U.S., so we should, too" (despite that fact that China is actually more advanced in many fields!). Indeed, declaring that the Black Ships have arrived is a good way to convince one's superiors of the need for action and a relatively simple way to persuade others of the importance of opening up the country.

However, we cannot wait for the Black Ships to arrive, and pursuing international cooperation is an important part of active manipulation from behind the scenes. When the story hinges on cooperatively sailing the Black Ships, we can find many ways of communicating with researchers overseas. International joint research projects provide a particularly easy way of obtaining funds; therefore, I believe they should be pursued to the extent possible.

The second approach I suggest is "Gundamization". Indeed, current smartphones empower human beings, making them similar to Gundam. We put on our Mobile Suits when we head out to face our enemies (whomever they may be). In particular, we should use tools such as machine translation and Skype as a way of heading out into the world and communicating with others. "Gundamization" is no longer a big deal.

Next, to promote system creation, it is necessary to always remember that systems are generally invisible and dynamic. It is easy to understand a new computer that runs one hundred times faster than the current one. It is more difficult to understand how a new system can solve problems one hundred times faster. Telling top executives what you want to do and why it is important is likely to be met with confusion due to the complexity of the issue. You will be told to come back with a simpler example, but having gone to the trouble to do so, you will then be told that what you propose is merely common sense. In the end, the first step in creating outstanding systems scientists is to teach field-specific knowledge. Another extremely important task is to learn what you do not understand from others.

6 Concluding Remarks

It is important that systems scientists understand their respective specialized areas and have a cross-disciplinary perspective for dealing with problems that arise. In addition, I believe we should use agent-based modeling. In conclusion, I introduce some statements slides that I frequently use in my lectures.

One is a statement that is engraved in the Picasso museum in Barcelona: "Art is a lie that makes us realize truth". The other is a book by Duncan Watts, *Everything Is Obvious: Once You Know the Answer* [4]. Slightly adjusting these two phrases provides a good description of how I think about complex systems and agent-based modeling:

- Agent simulation is the lie that helps us realize truth.
- Something will be obvious once you know the agent simulation.

I look forward to seeing future research on complex systems using agent modeling.

References

1. Terano, T.: Gallery for evolutionary computation and artificial intelligence researches: where do we come from and where shall we go. In: Kurahashi, S., et al. (eds.) Innovative Approaches in Agent-Based Modelling and Business Intelligence. Springer Agent-Based Social Science Series, vol. 12. Springer, Heidelberg (2018, to appear)
2. Kahneman, D.: Thinking, Fast and Slow. Penguin Books, London (2011)
3. Complexity Hub Vienna (ed.): 43 Visions in Complexity. World Scientific, New Jersey (2016)
4. Watts, D.: Everything is Obvious: Once You Know the Answer. Atlantic Books, London (2011)

Artificial Intelligence Technology and Social Problem Solving

Yeunbae Kim[1]([⊠]) and Jaehyuk Cha[2]

[1] Intelligent Information Technology Research Center, Hanyang University,
Seoul, Korea
kimyeunbae@hanyang.ac.kr
[2] Department of Computer Science, Hanyang University, Seoul, Korea
chajh@hanyang.ac.kr

Abstract. Modern societal issues occur in a broad spectrum with very high levels of complexity and challenges, many of which are becoming increasingly difficult to address without the aid of cutting-edge technology. To alleviate these social problems, the Korean government recently announced the implementation of mega-projects to solve low employment, population aging, low birth rate and social safety net problems by utilizing AI and ICBM (IoT, Cloud Computing, Big Data, Mobile) technologies. In this letter, we will present the views on how AI and ICT technologies can be applied to ease or solve social problems by sharing examples of research results from studies of social anxiety, environmental noise, mobility of the disabled, and problems in social safety. We will also describe how all these technologies, big data, methodologies and knowledge can be combined onto an open social informatics platform.

Keywords: Social problem solving · Artificial intelligence ·
Social informatics platform

1 Introduction

A string of breakthroughs in artificial intelligence has placed AI in increasingly visible positions in society, heralding its emergence as a viable, practical, and revolutionary technology. In recent years, we have witnessed IBM's Watson win first place in the American quiz show Jeopardy! and Google's AlphaGo beat the Go world champion, and in the very near future, self-driving cars are expected to become a common sight on every street. Such promising developments spur optimism for an exciting future produced by the integration of AI technology and human creativity.

AI technology has grown remarkably over the past decade. Countries around the world have invested heavily in AI technology research and development. Major corporations are also applying AI technology to social problem solving; notably, IBM is actively working on their Science for Social Good initiative. The initiative will build on the success of the company's noted AI program, Watson, which has helped address healthcare, education, and environmental challenges since its development. One particularly successful project used machine learning models to better understand the spread of the *Zika* virus. Using complex data, the team developed a predictive model

© Springer Nature Singapore Pte Ltd. 2019
F. Koch et al. (Eds.): GEAR 2018, CCIS 999, pp. 8–20, 2019.
https://doi.org/10.1007/978-981-13-6936-0_2

that identified which primate species should be targeted for *Zika* virus surveillance and management. The results of the project are now leading new testing in the field to help prevent the spread of the disease [1].

On the other hand, investments in technology are generally mostly used for industrial and service growth, while investments for positive social impact appear to be relatively small and passive. This passive attitude seems to reflect the influence of a given nation's politics and policies rather than the absence of technology.

For example, in 2017, only 4.2% of the total budget of the Korean government's R&D of ICT (Information and Communication Technology) was used for social problem solving, but this investment will be increased to 45% within the next five years as the improvement of Korean people's livelihoods and social problems are selected as important issues by the present government [2]. In addition, new categories within ICT, including AI, are required as a key means of improving quality of life and achieving population growth in this country.

In this letter, I introduce research on the informatics platform for social problem solving, specifically based on spatio-temporal data, conducted by Hanyang University and cooperating institutions. This research ultimately intends to develop informatics and convergent scientific methodologies that can explain, predict and deal with diverse social problems through a transdisciplinary convergence of social sciences, data science and AI. The research focuses on social problems that involve spatio-temporal information, and applies social scientific approaches and data-analytic methods on a pilot basis to explore basic research issues and the validity of the approaches. Furthermore, (1) open-source informatics using convergent-scientific methodology and models, and (2) the spatio-temporal data sets that are to be acquired in the midst of exploring social problems for potential resolution are developed.

In order to examine the applicability of the models and informatics platform in addressing a variety of social problems in the public as well as in private sectors, the following social problems are identified and chosen:

1. *Analysis of individual characteristics with suicidal impulse*
2. *Study on the mobility of the disabled using GPS data*
3. *Visualization of the distribution of anxiety using Social Network Services*
4. *Big data-based analysis of noise environment and exploration of technical and institutional solutions for its improvement*
5. *Analysis of the response governance regarding the Middle Eastern Respiratory Syndrome (MERS)*

The research issues in the above social problems are explored, and the validity of the convergent-scientific methodologies are tested. The feasibility for the potential resolution of the problems are also examined. The relevant data and information are stored in a knowledge base (KB), and at the same time research methods that are used in data extraction, collection, analysis and visualization are also developed. Furthermore, the KB and the method database are merged into an open informatics platform in order to be used in various research projects, business activities, and policy debates.

2 Pilot Research and Studies on Social Problem Solving

2.1 Analysis of Individual Characteristics with Suicidal Impulse

While suicide rates in OECD countries are declining, only South Korea has increasing suicide rates; moreover, Korea currently has the highest suicide rate among OECD countries as shown in Fig. 1. Its high suicide rate is one of Korea's biggest social problems, entailing the establishment of effective suicide prevention measures by understanding the causes of suicide. The goals of the research are to: (1) understand suicidal impulse by analyzing the characteristics of members of society according to suicidal impulse experience; (2) predict the likelihood of attempting suicide and ana-lyzing the spatio-temporal quality of life; and (3) to establish a policy to help prevent suicide.

The Korean Social Survey and Survey of Youth Health Status Data are used for the analysis of suicide risk groups through data mining techniques, using a predictive model based on cell propagation to overcome the limitations of existing statistics methods such as characterization or classification. In the case of the characterization technique, results indicate that there are too many features related to suicide, and that there are variables including many categorical values, making it difficult to identify the variables that affect suicide. On the other hand, the classification technique had diffi-culties identifying the variables that affect suicide because the number of members attempting suicide was too small.

Correlations between suicide impulses and individual attributes of members of society and the trends of the correlations by year are obtained. The concepts of *support*, *confidence* and *density* are introduced to identify risk groups of suicide attempts, and computational performance problems caused by excessive numbers of risk groups are solved by applying a *convex growing* method.

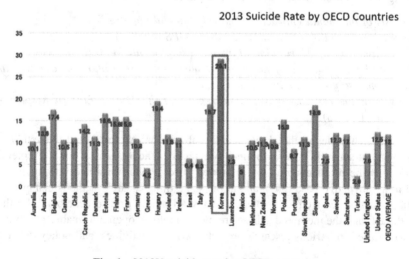

Fig. 1. 2013Y suicide rate by OECD countries

The 2014Y social survey including personal and household information of members of the society are used for analysis. The attributes include *gender, age, education, marital status, level of satisfaction, disability status, occupation status, housing,* and *household income.*

The high-risk suicide cluster was identified using a small number of convexes. A convex is a set of cells, with one cell being the smallest unit of the cluster for the analysis, and a density is the ratio of the number of non-empty cells to the total number of cells in convex C [3].

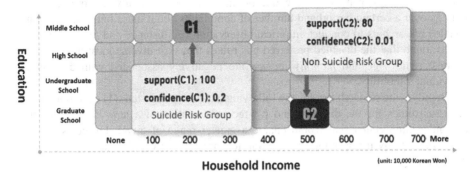

Fig. 2. Suicide risk groups represented by *household income* and *level of education*

Figure 2 shows that the highest suicidal risk group **C1** is composed of members with low income and education level. It was identified that level of *satisfaction* with life has the highest impact on suicidal impulse, followed in order of impact by *disability, marital status, housing, household income, occupation status, gender, age* and *level of education*. The results showed that women and young people tend to have more suicidal impulse.

New prediction models with other machine learning methods and the establishment of mitigation policies are still in development. Subjective analyses of change of well-being, social exclusion, and characteristics of spatio-temporal analysis will also be explored in the future.

2.2 Study on the Mobility of the Disabled Using GPS Data

Mobility rights are closely related to quality of life as a part of social rights. Therefore social efforts are needed to guarantee mobility rights to both the physically and mentally disabled. The goal of the study is to suggest a policy for the extension of mobility rights of the disabled. In order to achieve this, travel patterns and socio-demographic characteristics of the physically impaired with low levels of mobility are studied. The study focused on individuals with physical impairments as the initial test group as a means to eventually gain insight into the mobility of the wider disabled population. Conventional studies on mobility measurement obtained data from *travel diaries, interviews,* and *questionnaire surveys*. A few studies used geo-location tracking GPS data.

GPS data is collected via mobile device and used to analyze the mobility patterns (distance, speed, frequency of outings) by using regression analysis, and to search for methods to extend mobility. A new metrics for mobility with a new indicator (travel range) was developed, and the way mobility impacts the quality of life of the disabled has been verified [4].

About 100 people with physical disabilities participated and collected more than 100,000 geo-location data over a month using an open mobile application called *traccar*. Their trajectories are visualized based on the GPS data as shown in Fig. 3.

The use of location data explained mobility status better than the conventional questionnaire survey method. The questionnaire surveyed mainly the frequency of outings over a certain period and number of complaints about these outings. GPS data enabled researchers to conduct empirical observations on distance and range of travel. It was found that the disabled preferred bus routes that visit diverse locations over the shortest route. *Age* and *monthly income* are negatively associated with a disabled individual's mobility.

Based on the research results, the following has been suggested: (1) development of new bus routes for the disabled and (2) recommendation of a new location for the current welfare center that would enable a greater range of travel. Further study on travel patterns by using indoor positioning technology and CCTV image data will be deployed.

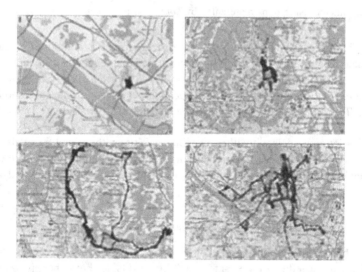

Fig. 3. Visualization of trajectory of disabled using geo-location data

2.3 Visualization of the Distribution of Anxiety Using Social Network Services

Many social issues including political polarization, competition in private education, increases in suicide rate, youth unemployment, low birth rate, and hate crime have anxiety as their background. The increase of social anxiety can intensify competition and conflict, which can interfere with social solidarity and cause a decrease in social trust.

Existing social science research mainly focused on grasping public opinion through questionnaires, and ignored the role of emotions. The Internet and social media were used to access emotional traits since they provide a platform not only for the active exchange of information, but also for the sharing and diffusion of emotional responses. If such emotional responses on the internet and geo-locations can be captured in real-time through machine learning, their spatio-temporal distribution could be visualized in order to observe their current status and changes by geographical region.

A visualization system was built to map the regional and temporal distribution of anxiety psychology by combining spatio-temporal information using SNS (Twitter) with sentiment analysis. A Twitter message collecting crawler was also developed to build a dictionary and tweet corpus. Based on these, an automatic classification system of anxiety-related messages was developed for the first time in Korea by applying machine learning to visualize the nationwide distribution of anxiety (See Fig. 4) [5].

An average of 5,500 tweets with *place_id* are collected using Open API *Twitter4j*. To date, about 820,000 units of data have been collected. A *Naïve Bayes* Classifier was used for anxiety identification. An accuracy of 84.8% was obtained by using 1,750 and 70,830 anxiety and non-anxiety tweets as training data respectively, and 585 and 23,610 anxiety and non-anxiety tweets as testing data, respectively.

The system indicated the existence of regional disparities in anxiety emotions. It was found that Twitter users who reside in politicized regions have a lower degree of disclosure about their residing areas. This can be interpreted as the act of avoiding situations where the individual and the political position of the region coincide.

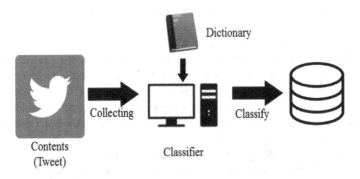

Fig. 4. Process of Twitter message classification

As anxiety is not a permanent characteristic of an individual, it can change depending on the time and situation, making it difficult to measure by questionnaire survey at any given time. The Twitter-based system can compensate for the limitations of such a survey method because it can continuously classify accumulated tweet text data and provide a temporal visualization of anxiety distribution at a given time within a desired visual scale (by ward, city, province and nationwide) as shown in Fig. 5.

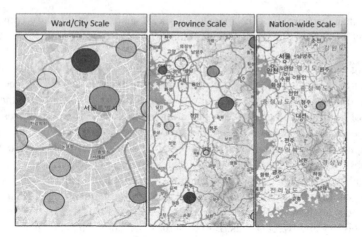

Fig. 5. Regional distribution of anxiety in Korean society and visualization by geo-scale

2.4   Big Data-Based Analysis of Noise Environment and Exploration of Technical and Institutional Solutions for Its Improvement

Environmental issues are a major social concern in our age, and interest has been increasing not only in the consequences of pollution but also in the effects of general environmental aesthetics on quality of life. There is much active effort to improve the visual environment, but not nearly as much interest has been given to improve the auditory environment. Until now, policies on the auditory environment have remained passive countermeasures to simply quantified acoustic qualities (e.g., volume in dB) in specific places such as construction sites, railroads, highways, and residential areas. They lack a comprehensive study of contextual correlations, such as the physical properties of sound, the environmental factors in time and space, and the human emotional response of noise perception.

The goal of this study is to provide a cognitive-based, human-friendly solution to improve noise problems. In order to achieve this, the study aimed to (1) develop a tool for collecting sound data and converting into a sound database, and (2) build spatio-temporal features and a management platform for indoor and outdoor noise sources.

First, pilot experiments were conducted to predict the indicators that measure emotional reactions by developing a handheld device application for data collection.

Three separate free-walking experiments and in-depth interviews were conducted with 78 subjects at international airport lobbies and outdoor environments.

Through the experiment, the behavior patterns of the subjects in various acoustic environments were analyzed, and indicators of emotional reactions were identified. It was determined that the psychological state and the personal environment of the subject are important indicators of the perception of the auditory environment. In order to take into account both the psychological state of the subject and the physical properties of the external sound stimulus, an omnidirectional microphone is used to record the entire acoustic environment.

118 subjects with smartphones with the built-in application walked for an hour in downtown Seoul for data collection. On the app, after entering the prerequisite information, subjects pressed '*Good*' or '*Bad*' whenever they heard a sound that caught their attention. Pressing the button would record the sound for 15 s, and subjects were additionally asked to answer a series of questions about the physical characteristics of the specific location and the characteristics of the auditory environment. During the one-hour experiment, about 600 sound environment reports were accumulated, with one subject reporting the sound characteristics from an average of 5 different places.

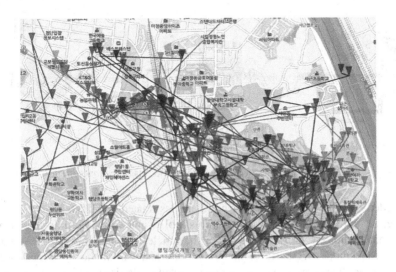

Fig. 6. Subject's paths and marks for sound types

Unlike previous studies, the subjects' paths were not pre-determined, and the position, sound and emotional response of the subject are collected simultaneously. The paths can be displayed to analyze the relations of the soundscapes to the paths (Fig. 6).

The study helped to build a positive auditory environment for specific places, to provide policy data for noise regulation and positive auditory environments, to identify the contexts and areas that are alienated from the auditory environment, and to extend the social meaning of "noise" within the study of sound.

2.5   Analysis of the Response Governance Regarding the Middle Eastern Respiratory Syndrome (MERS)

The development and spread of new infectious diseases are increasing due to the expansion of international exchange. As can be seen from the MERS outbreak in Korea in 2015, epidemics have profound social and economic impacts. It is imperative to establish an effective shelter and rapid response system (RRS) for infectious diseases control.

The goal of the study is to compare the official response system with the actual response system in order to understand the institutional mechanism of the epidemic response system, and to find effective policy alternatives through the collaboration of policy scholars and data scientists.

Web-based newspaper articles were analyzed to compare the official crisis response system designed to operate in outbreaks to the actual crisis response. An automatic news article crawling tool was developed, and 53,415 MERS-related articles were collected, clustered and stored in the database (Fig. 7).

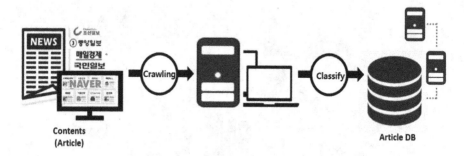

Fig. 7. Automatic news article collection & classification system

In order to manage and search for news articles related to MERS from the article database, a curation tool was developed. This tool is able to extract information into triplet graphs (subjects/verbs/objects) from the articles by applying natural language processing techniques. A basic dictionary for the analysis of the infectious disease response system was created based on the extracted triplet information. The information extracted by the curation tool is massive and complex, which limits the ability to correctly understand and interpret information.

A tool for visualizing information at a specific time with a network graph was developed and utilized to facilitate analysis and visualization of the networks (Fig. 8). All tools are integrated into a single platform to maximize the efficiency of the process.

Fig. 8. Visualization of graph network by specific time

As for the official crisis response manual in case of an infectious disease, social network analysis indicated that while the National Security Bureau (NSB) and Public Health Centers play as large a role as the Center for Disease Control (CDC) in crisis management, the analysis of the news articles showed that the NSB was in fact rarely mentioned. It was found that the CDC and Central Disaster Response Headquarters, the official government organizations that deal with infectious diseases, as well as the Central MERS Management Countermeasures & Support Headquarters, a temporarily established organization, were not playing an important role in response to the MERS outbreak. On the other hand, the Ministry of Health and Welfare, medical institutions, and local governments all have played a central role in responding to MERS. This means that the structure and characteristics of the Command & Control and communication in the official response system seems to have a decisive influence on the cooperative response in a real crisis response. These results provided concrete information on the role of each respondent and the communication system that previous studies based on interviews and surveys have not found.

Much research based on machine learning has been criticized for giving more importance on method itself from the start rather than focusing on data reliability.

This study is based on a KB in which policy researchers manually analyze news articles and prepare basic data by tagging them. This way, it provides a basis for improving the reliability of results when executing text mining work through machine learning.

By using text mining techniques and social network analysis, it is possible to get a comprehensive view of social problems such as the occurrence of infectious diseases by examining the structure and characteristics of the response system from a holistic perspective of the entire system.

With the results of this study, new policies for infectious disease control are suggested in the following directions: (1) Strengthen cooperation networks in early response systems of infectious diseases; (2) Develop new, effective and efficient management plans of cooperative networks; and (3) Create new research to cover other diseases such as avian influenza and SARS [6].

3   Convergent Approaches and Open Informatics Platform

An ever-present obstacle in the traditional social sciences when addressing social issues are the difficulties of obtaining evidences from massive data for hypothesis and theory verification. Data science and AI can ease such difficulties and support social science by discovering hidden contexts and new patterns of social phenomena via low-cost analyses of large data. On the other hand, knowledge and patterns derived by machine learning from a large data set with noise often lack validity. Although data-driven inductive methods are effective for finding patterns and correlations, there is a clear limitation to discovering causal relationships.

Social science can help data science and AI by interpreting social phenomena through humanistic literacy and social-scientific thought to verify theoretical validity, and identifying causal relationships through deductive and qualitative approaches. This is why we need convergent-scientific approaches for social problem solving.

Convergent approaches offer the new possibility of building an informatics platform that can interpret, predict and solve various social problems through the combination of social science and data science.

In all 5 pilot studies, the convergent-scientific approaches are found valid and sound. Most of the research agendas involved the real-time collection and development of spatio-temporal databases in a real-time manner, and analytic visualization of the results. Such visualization promises new possibilities in data interpretation. The data sets and tools for data collection, analysis and visualization are integrated onto an informatics platform so that they can be used in future research projects and policy debates.

The research was the first transdisciplinary attempt to converge social sciences and data sciences in Korea. This approach will offer a breakthrough in predicting, preventing and addressing future social problems. The research methodology, as a trailblazer, will offer new ground for a research field of a transdisciplinary nature converging data sciences, AI and social sciences. The data, information, knowledge, methodologies, and techniques will all be combined onto an open informatics platform. The platform will be maintained on an open-source basis so that it can be used as a hub for various academic research projects, business activities, and policy debates (See Fig. 9). The Open Informatics Platform is planned to be expanded to incorporate citizen sensing, in which people's observations and views are shared via mobile devices and Internet services in the future [7].

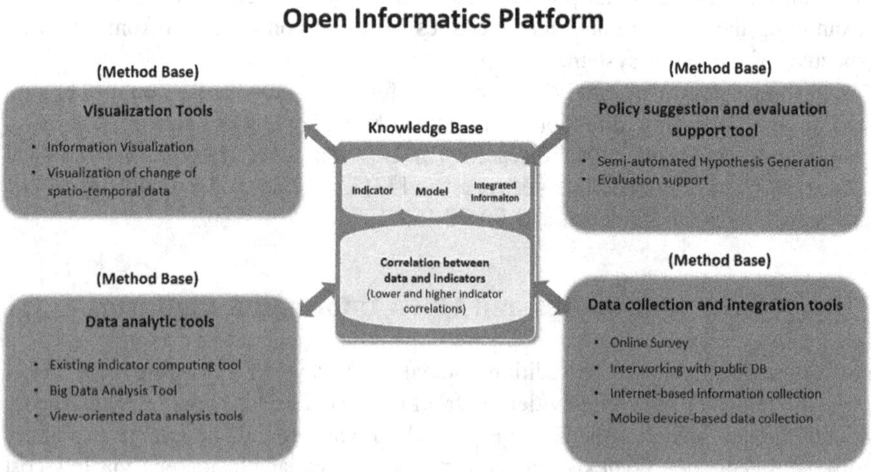

Open Informatics Platform

Fig. 9. Structure of informatics platform

4 Conclusions

In the area of social problem solving, fundamental problems have complex political, social and economic aspects that have their roots in human nature. Both technical and social approaches are essential for tackling social problem solving. In fact, it is the integrated, orchestrated marriage between the two that would bring us closer to effective social problem management.

We need to first study and carefully define the indicators specific to a given social problem or domain. There are many qualitative indicators that cannot be directly and explicitly measured such as social emotions, basic human needs and rights, and life fulfillment [8].

If the results of machine learning are difficult to measure or include combinations of results that are difficult to define, that particular social problem may not be suitable for machine learning. Therefore, there is a need for new social methods and algorithms that can accurately collect and identify the measurable indicators from opinions of social demanders. Recently, MIT has developed a device to quantitatively measure social signals. The small, lightweight wearable device contains sensors that record the people's behaviors (physical activity, gestures, and the amount of variation in speech prosody, etc.) [9].

Machine learning technologies working on already existing data sets are relatively inexpensive compared to conventional million-dollar social programs since machine learning tools can be easily extended. However, they can introduce bias and errors depending on the data content used to train machine learning models or can also be misinterpreted. Human experts are always needed to recognize and correct erroneous outputs and interpretations in order to prevent prejudices [10].

In the development of AI applications, a great amount of time and resources are required to sort, identify and refine data to provide massive data for training. For instance, machine learning models need to learn millions of photos to recognize specific animals or faces, but human intelligence is able to recognize visual cues by looking at only a few photos. Perhaps it is time to develop a new AI framework which can infer and recognize objects based on small amounts of data, such as Transfer Learning [11], generate lacking data (GAN), or integrate traditional AI technologies, such as symbolic AI and statistical machine learning into new frameworks.

Machine learning is excellent in predicting, but many social problem solutions do not depend on predictions. The organic ways solutions to specific problems actually unfold according to new policies and programs can be more practical and worth studying than building a cure-all machine learning algorithm. While the evolution of AI is progressing at a stunning rate, there are still challenges to solving social problems. Further research on the integration of social science and AI is required.

A world in which artificial intelligence actually makes policy decisions is still hard to imagine. Considering the current limitations and capabilities of AI, AI should primarily be used as a decision aid.

Acknowledgements. This work was supported by the National Research Foundation of Korea (NRF) grant funded by the Korean government (MSIT[1]) (No. 2018R1A5A7059549).

---

[1] Ministry of Science and ICT.

References

1. IBM: Science for Social Good – Applying AI, cloud and deep science toward new societal challenges. http://www.research.ibm.com/science-for-social-good/
2. MSIT of Korea: I-KOREA 4.0 ICT R&D Innovation Strategy (2018)
3. Kim, N., Hong, J., Kim, H., Kim, S.: Analyzing suicide-ideation survey to identify high-risk groups: a data mining approach. In: International Conference on Green and Human Information, China (2017)
4. Kim, H., Lee, Y., Cha, J.: Mobility among people with physical impairment: a study using geo-location tracking data. In: International Conference on Green and Human Information, China (2017)
5. Lee, J., Kim, J., Choi, Y.: SNS data visualization for analyzing spatial-temporal distribution of social anxiety. In: EDB 2016 Proceedings of the Sixth International Conference on Emerging Databases: Technologies, Applications, and Theory, Jeju, Korea, pp. 106–109 (2016)
6. Kim, J., Jung, J., Cha, J., Choi, J., Choi, C., Oh, S.: Application of network analysis into emergency response: focusing on the 2015 outbreak of the middle-eastern respiratory syndrome in Korea. In: Information, vol. 21, No. 2, pp. 441–446. International Information Institute, Tokyo (2018)
7. Koch, F., Cardonha, C., Gentil, J.M., Borger, S.: A platform for citizen sensing in sentient cities. In: Nin, J., Villatoro, D. (eds.) CitiSens 2012. LNCS (LNAI), vol. 7685, pp. 57–66. Springer, Heidelberg (2013). https://doi.org/10.1007/978-3-642-36074-9_6
8. Land, K.C., Ferriss, A.L.: Chapter 52: The sociology of social indicators. In: 21st Century Sociology (2006). http://www.soc.duke.edu/~cwi/Section_I/I-19TheSociologyofSocial Indicators.pdf
9. Pentland, A.: To signal is human. Am. Sci. **98**(3), 204–211 (2010)
10. Kleinberg, J., Ludwig, J., Mullainathan, S.: A guide to solving social problems with machine learning. In: Harvard Business Review (2016). https://hbr.org/2016/12/a-guide-to-solving-social-problems-with-machine-learning
11. Yosinski, J., Clune, J., Bengio, Y., Lipson, H.: How transferable are features in deep neural networks? In: Advances in Neural Information Processing Systems 27 (NIPS 2014). NIPS Foundation (2014)

A Formal Model of Managerial Decision Making for Business Case Description

Masaaki Kunigami[1($\boxtimes$)] (ID), Takamasa Kikuchi[2] (ID),
and Takao Terano[3] (ID)

[1] Tokyo Institute of Technology, Yokohama, Japan
mkunigami@gakushikai.jp
[2] Keio University, Yokohama, Japan
[3] Chiba University of Commerce, Ichikawa, Japan

Abstract. This paper proposes a novel formal description model of organizational decision-making: the Managerial Decision-Making Description Model (MDDM). This model introduces a four terminal element representation to describe managerial decisions redefining relationship between their objectives and resources. It enables them to compare various decision-making processes from not only actual business cases but also virtual ones, from an agent-based simulation, too. This model is also applicable in facilitation support for business gaming.

Keywords: Decision making · Formal description · Business case

1 Introduction

This paper proposes a formal description model to describe managerial decision-making processes used to transform the business organization. This heuristic tool named as the Managerial Decision-making Description Model (MDDM) provides common way to visualize a decision-making process within the business case method as well as the agent-based organizational simulation and business gaming.

In contrast to the Case Management Model and Notation (OMG [6]) with the Business Process Model and Notation (OMG [4]) and the Decision Model and Notation (OMG [5]) describing the static process of the business, MDDM focuses on describing the organizational decision-making that causes a one-shot transition process in which the whole business structure is changed.

A High-Level Business Case in Sawatani [7] is presented to describe such a one-shot transition of the business structure. While their HLBC represents the evolution of the functions and services of the business structure, the MDDM focuses on the decision-making process driving the transition of the business structure.

Accordingly, we started by defining the key terminologies for the MDDM. First, the business structure of the organization is defined as a multi-layered structure of combinations of business objectives and the related resources or means. Next, managerial decision-making is understood as a means that an agent (i.e., an actor or a player in the organization) defines or redefines the combinations of business objectives and their related resources in the business structure.

© Springer Nature Singapore Pte Ltd. 2019
F. Koch et al. (Eds.): GEAR 2018, CCIS 999, pp. 21–26, 2019.
https://doi.org/10.1007/978-981-13-6936-0_3

To formally describe the managerial decision-making that change the business structure of the organization, we require that MDDM represent following items.

(a) the multi-layered structure of the organizational business, and its transition,
(b) the focus (or bounded scope) of the agent's observations and actions,
(c) the agent's position corresponding to each layer in the business structure,
(d) the chronological order and the causality of the agents' decisions.

Those enables the MDDM to describe "who" decides "what", "when," and "where it fall within the business structure," along with how the decisions change.

2 Methodology

To represent the transition of the business structure as a decision diagram, the Managerial Decision-making Description Model (MDDM) uses three kinds of components. With placing and connecting those components, the decision diagram describes organizational decision-making as if it were an equivalent circuit. The decision diagram satisfies the condition presented in chapter one.

2.1 Three Major Components

The MDDM uses following three kinds of major components: (i) the business structure component, (ii) the environment component and (iii) the agents' decision element.

• Business Structure Component

This component represents the multi-layered structure of objectives-resources couplings in the organizational business process. It comprises the objective symbols, resource symbols and the connections between them. Each objective symbol represents a goal, an objective, or a target in the layer of business. A resource symbol represents a resource, an operation, a product or a means required to achieve the objective symbol. By heaping up the objectives-resources couplings, the business structure component represents the multi-layered structure of the business.

• Environment Component

This component describes the states, the transitions of the status, and events outside of the organization. It consists of status and event symbols. Each status represents the situation or condition of technology, the market or another organization. The event indicates something that happens with the status and triggers the agent's decision, or indicates something that is caused by the agent's decision. The order of these symbols from left to right indicates their chronological order.

• Agents' Decision Element

The agent's decision element describes how the agent redefines the objectives and means in the business structure of the organization. Each decision of the agents is represented as a "decision element" with 2×2 terminals.

Each terminal has a specific function. The left hand's dual terminals of a decision element represent the agent's observation-action pair before the decision. In contrast, the right-hand two terminals represent the agent's observation-action pair as a consequence of the agent's decision. The upper dual terminals indicate the agent's objectives or targets, and the lower dual terminals indicate the corresponding resources or means to facilitate those objectives.

2.2 Composing the Decision Diagram

With allocating and connecting those components, the MDDM describes the managerial decision-making as a decision diagram (Fig. 1).

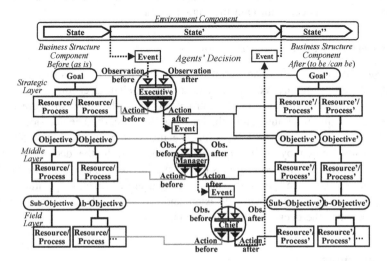

Fig. 1. The MDDM represents the managerial decision-making as a decision diagram with the three components, the Environment (top), the business structure (right and left side) and the agent's decision (four terminal elements between the Business Structures).

To begin with, the environment component is placed at the top or bottom of the decision diagram. It introduces the time procession (from old to new) from a horizontal direction (from left to right) of the decision diagram.

Next, to describe the transition of the business structure, the two business structure components are placed on the left and right sides, respectively. The left-side component represents the business structure that existed before the agents' decision. The right-side one represents the consequences of the agents' decisions. We call the left-side structure "before" or "as is", and call the right-side one "after", "to be" or "outcome". These business structures introduce vertical layers into the decision diagram from strategic management (upper) to the field operations (lower).

Third, the agent's decision elements are allocated between the business structures. The allocation of the decision elements reflects the organizational position and chronological order of the agents' decisions.

Fourth, these agent's decision elements are connected to the other components and decision elements. Each upper left terminal is connected to the symbols that the agent had observed as the objective or the target in the left ("before") business structure component. Each lower left terminal is connected to the symbols that the agent had taken action to as the resources or the means in the left ("before") business structure. Similarly, each upper right and lower right terminals are connected to the new objective and resources symbols in the right ("after") business structure respectively.

Finally, an environment-agent interaction or agent-agent interaction are represented by connecting the agent's terminals and related event symbols. For example, when an event that is related to the environment triggers the agent's decision, the event is connected to the agent's upper left terminal. Similarly, if an agent's decision triggers another agent's decision, the agent's lower right terminal and the other agent's upper right terminal are connected through the trigger event.

2.3 Properties of the Decision Diagram

The decision diagram of the managerial decision-making enables us to describe the following properties that are required in chapter one. (a) The decision diagram represents the multi-layered structure that is introduced by the business structure components before and after the transition of the business structure. (b) The decision diagram represents that each agent decides with specific observation-action (objectives-resources) pairs limited by their scope and position. (c) In the decision diagram, each agent's vertical position corresponds to the layer of the business structure to which the agent belongs. (d) In the decision diagram, each agent's horizontal position reflects the chronological order of the agents' decision, and event symbol connections represent causalities between the decisions and the events.

3 Applications

Here we briefly illustrate how the MDDM discriminates among some typical managerial decision-making styles. First, we show that the decision diagram reflects the difference among top-down style, bottom-up style and a style tolerant of informal communication. Next, we show another example of the decision diagram representing the differences and similarities of the KAIZEN activity and organizational deviation.

3.1 Differences in Decision-Making Styles

The MDDM discriminates between decision-making styles by allocating the decision elements both in the vertical layers and the horizontal chronological order.

As the most fundamental pattern, MDDM discriminates between top-down and bottom-up style of the managerial decision-making. In the decision diagram, the top-down style is described as the allocation of the decision elements from the upper-left (i.e., the strategic decision comes first) to the lower-right (i.e., the field-level decisions follow). In contrast, the bottom-up style decision-making is described by the decision elements' allocation from the lower-left (i.e., the field-level decision comes first) to the upper-right (i.e., the strategic decision follows) (Fig. 2).

Next, the MDDM describes the typical pattern of the informal communication making a shortcut in the organizational hierarchy. Related study Toriyama [8] showed that such informal communication promotes organizational decision-making. The decision diagram indicates the existence, related agents, content and point in time of the informal communication (Fig. 2).

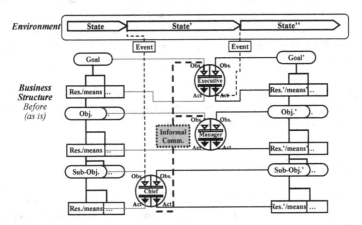

Fig. 2. A bottom-up decision making style is represented by the decision elements from the lower-left to the upper-right. An informal communication is expressed with the event (dotted symbol) among related agents' decision and the connections between them (thick broken lines).

3.2   Similarities in Decision-Making and Different Outcomes

MDDM describes both the similarities and differences of managerial decision-making. The related literature Kobayashi [2] mentioned that the spontaneous innovation and organizational deviation was derived by a common mechanism but caused the opposite consequences in terms of social aspects. They also demonstrated that the written cases from their simulation outcomes were grounded in the actual business case. MDDM illustrates formally the similarities and differences.

In such cases, MDDM represents the common mechanism by using a similar allocation within the decision diagram, and introduces an auxiliary business structure representing the social expectations on the right-end. The decision diagram discriminates between the spontaneous innovation and the organizational deviation from the aspect whether the "after" business structure is better or worse than the socially expected business structure.

4   Summary and Remarks

In summary, propose the formal description model for the business case. MDDM provides the decision diagram that illustrates the transition of the business structure caused by the related agents' decisions. The decision diagram also represents the chronological order and causalities between the decisions themselves and the decisions and the environment. MDDM discriminates between the decision style in the business cases, e.g., top-down, bottom-up, or informal communication.

As noted in the paper, MDDM provides decision diagrams that result from the transformation of formal descriptions of organizational, agent-based simulation (ABS) logs, along with business game logs as well as of the actual business cases. The case of the Kaizen and deviation, found in Kobayashi [1, 2] indicates that there is no essential difference between cases from organizational ABS logs and actual business cases. Because of limited space, we will exemplify the business simulation analysis with MDDM in forthcoming paper being prepared. Nakano [3] has already presented the business simulation gaming integrated with case learning, based on actual business cases. The MDDM will provide an effective way to describe the players' decisions and to compare them formally to the original business case.

Acknowledgments. The authors express sincere appreciation for the authors of the work cited in this paper and to the anonymous reviewers of the GEAR2018 conference. The authors would like to thank Enago for the English language review.

References

1. Kobayashi, T., Takahashi, S., Kunigami, M., Yoshikawa, A., Terano, T.: A unified agent-based model to analyze organizational deviation and kaizen activities. In: Dechesne, F., Hattori, H., ter Mors, A., Such, J.M., Weyns, D., Dignum, F. (eds.) AAMAS 2011. LNCS (LNAI), vol. 7068, pp. 384–395. Springer, Heidelberg (2012). https://doi.org/10.1007/978-3-642-27216-5_29
2. Kobayashi, T., Takahashi, S., Kunigami, M., Yoshikawa, A., Terano, T.: Is there innovation or deviation? analyzing emergent organizational behaviors through an agent based model and a case design. In: Proceedings on the 5th International Conference on Information, Process, and Knowledge Management (eKNOW 2013), pp. 166–171 (2013)
3. Nakano, K., Matsuyama, S., Terano, T.: Research on a learning system toward integration of case method and business gaming. In: Proceedings on the 4th International Workshop on Agent-based Approach in Economic and Social Complex Systems (AESCS 2007), pp. 21–32 (2007)
4. Object Management Group: The Business Process Model and Notation Specification ver. 2.0.2, January 2014. https://www.omg.org/spec/BPMN/
5. Object Management Group: The Decision Model and Notation Specification ver. 1.2, March 2016. https://www.omg.org/spec/DMN
6. Object Management Group: The Case Management Model and Notation Specification ver. 1.1, December 2016. https://www.omg.org/spec/CMMN/
7. Sawatani, Y., Kashino, T., Goto, M.: Analysis and Findings on Innovation Creation Methodologies (2016). https://www.slideshare.net/YurikoSawatani/analysis-and-findings-on-innovation-creation-methodologies, slide 15. Accessed 23 June 2018
8. Toriyama, M., Kikuchi, T., Yang, C., Yamada, T., Terano, T.: Who is a key person to transfer knowledge. In: Proceedings on the 5th Knowledge Management in Organizations (KMO 2010), pp. 41–51 (2010)

Evolutionary Computation Meets Multiagent Systems for Better Solving Optimization Problems

Vinicius Renan de Carvalho$^{(\boxtimes)}$ and Jaime Simão Sichman

Laboratório de Técnicas Inteligentes (LTI), Escola Politécnica (EP),
University of Sao Paulo (USP), São Paulo, Brazil
{vrcarvalho,jaime.sichman}@usp.br

Abstract. In this work, we discuss the synergy between Evolutionary Computation (EC) and Multi-Agent Systems (MAS) when both are used together to enhance the process of solving optimization problems. Evolutionary algorithms are inspired by nature and follow Darwin theory of the fittest. They are usually applied where there is no specific algorithm which can solve optimization problems in a reasonable time. Multi-Agent Systems, in their turn, are collections of autonomous entities, named agents, that sense their environment and execute some actions in the environment to meet their individual or common goals. When these two techniques are applied together, one can create powerful approaches to better solve optimization problems. This paper presents an overview of this combined approach, considering both mono-objective and multi-objective approaches. In particular, we stress the importance of hyper-heuristic approaches, i.e., heuristics that help to choose the best EC algorithm among a candidate set.

Keywords: Hyper-heuristics ·
Multi-objective Evolutionary Algorithms · Voting methods ·
Agent cooperation

1 Introduction

An *optimization problem* is the problem of finding the best possible solution among a set of candidate solutions. More precisely, it aims to find a solution in the feasible region that minimizes (or maximizes) the value of a certain objective function [2]. In an optimization problem, solutions are composed of a set of decision variables $s = (x_1, \ldots, x_n)$, whose values belong to a set of domains $D = (D_1, \ldots, D_n)$. These domains can be either continuous ($D = \mathbb{R}$) or discrete ($D = \mathbb{Z}$). Thus, solutions can solve discrete or continuous optimization problems. In order to determine which solution better solves a given optimization problem, it is necessary to define a *fitness function* $f(s)$ that evaluates the quality of each solution.

© Springer Nature Singapore Pte Ltd. 2019
F. Koch et al. (Eds.): GEAR 2018, CCIS 999, pp. 27–41, 2019.
https://doi.org/10.1007/978-981-13-6936-0_4

However, many real-world problems are multi-criteria, and thus need the specification of multiple fitness functions in order to evaluate their solutions. This is the case, for example, of buying a car considering price and fuel consumption. These problems are called *Multi-Objective Problems* (MOPs), where the solutions should optimize different and often conflicting criteria [16]. Usually, classical exact optimization methods cannot be used to deal with MOPs and more sophisticated techniques are required. In this paper, evolutionary algorithms and hyper-heuristics are addressed to solve both mono-objective and multi-objective optimization problems.

2   Evolutionary Computation

According to Pearl [38], *heuristics* can be defined as criteria, methods or principles for deciding which among several alternatives courses of action promises to be the most effective in order to achieve some goal. Heuristics do not guarantee optimal solutions; in fact, they do not guarantee any solutions at all: all that can be said for a useful heuristic is that it offers solutions which are good enough most of the time [20].

Meta-heuristics, in turn, can be defined as an iterative generation process which guides a subordinate heuristic by combining intelligently different concepts for exploring and exploiting the search space, using strategies to structure information in order to find efficiently near-optimal solutions [35]. Usually, heuristics are specialized in solving problems for one particular domain, while meta-heuristics are more generic and adaptive in several domains.

One of the most used meta-heuristics are classified as *Evolutionary Computation* (EC) algorithms. It is the general term for several optimization algorithms that are inspired by the Darwinian principles of nature's capability to evolve living beings well adapted to their environment [7]. These algorithms are also called as *Evolutionary Algorithms* (EA), and they all share a common underlying idea of simulating the evolution of individual (or solution) structures via processes of selection, recombination, and mutation reproduction, thereby producing better solutions [7].

2.1   Evolutionary Algorithms

In the literature, we can find some algorithms that implement the concept of an evolutionary algorithm. That is the case of *Genetic Algorithm* [22] and of *Genetic Programming* [25]. In a genetic algorithm, individuals from the population compete and generate offspring using crossover and mutation. Genetic Programming employs more complex data representation than GA, such as a tree, to represent individuals. Thus, allowing individuals to have different lengths.

Both algorithms are focused on mono-objective optimization, that means, one single value to represent the quality of a given solution. However, several real-world problems considerate more than one fitness value in order to properly evaluate an individual quality. In this scenario, Multi-objective Evolutionary Algorithms (MOEAs) are able to find good solutions for this kind of problems [16].

Choosing an algorithm to solve a particular optimization problem is not a trivial task. Usually, a tuning method is necessary, if there is no previous knowledge about which algorithm to use and what is the recommended algorithm configuration to solve a given optimization problem. The tuning process consists in solving an optimization problem using different algorithm running instances (where each instance represent a different configuration), taking their results, and finding the best according to a quality indicator. The tuning task has to be executed for several times, since meta-heuristics are not deterministic algorithms. Some research has been proposed in order to reduce this difficult task. That is the case of *hyper-heuristics* [8].

2.2 Hyper-Heuristics

The motivation for proposing hyper-heuristic came from the "No Free Lunch Theorem" which establish that "for any algorithm, any elevated performance over one class of problems is an offset by diminished performance over another class" [45].

Hence, hyper-heuristics are defined as a high-level methodology that—given a particular problem instance or class of instances, and some low-level heuristics (LLH), or components—automatically produces an adequate combination of them to solve the problem efficiently [8].

Most of the research focuses on the selection of online hyper-heuristics, meaning that the process of finding solutions for the optimization problem tries to figure out which LLH should be given more time to execute. These works usually consider heuristics such as differential evolution, crossover, and mutation as low-level heuristics (LLH). However, there are some works which consider the whole algorithms as LLHs. The majority of research in this area has been limited to focus on single-objective optimization problems [27].

Evolutionary algorithms and hyper-heuristics have been developed along the years, most of the times considering centralized implementations. Given their number of complex components, they could be built as a multi-agent system. In the next section, we analyze how these areas have been applied together.

3 Multi-Agent Systems

According to Wooldridge [46], an agent is a computer system that is situated in some environment, and that is capable of autonomous action in this environment in order to meet its delegated objectives. An agent perceives the environment by sensors. Thus, based on his perceptions and considering his internal knowledge and beliefs, the agent can plan how to act, that means, using his actuators in order to update the environment. A multi-agent system (MAS) is one that consists of a number of agents, which interact with one another, typically by exchanging messages through some computer network infrastructure [46].

4 Combining Multi-Agent Systems and Evolutionary Computation

Evolutionary algorithms can be used by an agent in order to perform some tasks, such as learning, parameter estimations, or to support coordination of some group (team) activity, e.g. planning [39]. On the other hand, evolutionary algorithms can be built using the MAS background, which can provide the mechanisms for a decentralized search.

4.1 Multi-Agent Systems and Evolutionary Algorithms

Several approaches have been proposed for combining MAS and Evolutionary Algorithms. We can classify them as the following:

- Solutions as passive agents: in this case, solutions may act autonomously for instance, by selecting which partner would be more interesting to generate new offspring. In this category, agents are passive and do not perform any operation, but the algorithm perform operations;
- Solutions as active agents: in this case, solutions are agents which actively can generate new offspring. In this case, agents contain operations such as crossover and mutation;
- Algorithms or algorithm components as agents: in this case, complete algorithms instances are considered to generate hybrid approaches where algorithms agents share their solutions along the search;
- Multi-objective: to identify which approach deals with multi-objective optimization;
- Specially Organized: means if agents are organized in a specific structure, such as lattice and neighborhood.

Solutions as Passive Agents: In most research, agents represent a complete feasible solution (an individual in the population) to a given optimization problem. Thus, agents can cooperate (usually by crossover reproduction), compete for survival, observe and communicate with the environment. In this case, instead of the centralized view of a population of solutions, these works consider a population of agents, each one representing a solution. Using this concept, Eiben et al. [19] solved Travel Salesman Person problem by implementing a blackboard system which allowed the interchange among agents. Thus, each agent could access other agent's current solution by means of the blackboard mechanism.

In [39], solution agents also considered a non-renewable resource named life energy, which is increased or decreased based on how well an agent solve the given problem. This kind of agent representation was also applied in [47] for dynamic optimization problems, where the goal of an algorithm is no longer to find an optimal solution but to track the moving optima in the search. In [15] propose an approach to solve multi-objective problems: in this approach, each agent is invited to be a mate and can accept or decline the proposal according to its own strategy. These agents can be part of a society of agents, and can only accept agents from the same society as mates. Offspring are assigned to a certain society according to a dominance concept. In [43], the authors present an agent-based memetic algorithm for solving nonlinear optimization problems with equality constraints, where agents can be either cooperative or competitive through the simulated binary crossover or a proposed Life Span Learning Process. Chalupa [13] proposes a Multi-agent evolutionary algorithm (MEA). In this approach, each agent has its lifespan assigned when the solution is created. MEA performs many short-term local search subroutines, instead of long-term local search. After that, a collection of promising solutions are stored in the elite list. Thus, MEA relies on a well-tuned process of highly organized restarts from promising solutions and performing short-term local search subroutines to improve these solutions. In [44] the knowledge-based multi-agent evolutionary algorithm (KMEA) is proposed for solving the semiconductor final testing scheduling problem. KMEA has two phases: mutual-learning and competition. In the mutual-learning, each agent learns from the best one of its neighbors to obtain a better solution. In the competition, each agent competes with its current best neighbor. If it loses the competition, it will be replaced by a new agent generated according to the knowledge base.

Some research defined how solution agents should be set in the agent community. That is the case of [6], where agents reside in a grid, and each cell in the grid contains one agent. A similar approach was employed by Sun and Zhou [40], where the authors found solutions for the multi-objective energy resource scheduling of micro-grid, by considering a micro-grid as agents. In [37], all agents live in a lattice-like environment and die when the energy finishes. They can only interact with their neighbors. The local environments of all the agents are constructed by a social acquaintance net. Zeng et al. [48] also set agents in a lattice in order to find solutions for the Assembly Sequence Planning.

Solutions as Active Agents: Like in mentioned researches, some studies treat solution as agents, but they also consider the operator which generate the solution as a component. That is the case of [30], where agents have additional capacities of decision, learning, and cooperation, by using several operators which are scheduled by an adaptive decision process. The decision rules of the agents are adapted during the optimization process by reinforcement learning and mimetism. In [24] proposed a novel multi-agent multi-objective evolutionary framework based on trust where each solution is represented as an intelligent agent, and evolutionary operators and control parameters are represented as services. Agents select services in each generation based on trust that measures the competency or suitability of the services for solving particular problems. Huang et al. [23] propose an approach to solve the software module clustering problem (SMCP). They designed three evolutionary operators as agents: neighborhood competition operator, the mutation operator, and the self-learning operator. They also employed a similar energy concept used in earlier works. In [26] they solved the 3-D protein structure prediction using an approach which considers crossover and swap operations as agents. Each agent work with the complete protein (torsional angles) representation, called solutions, like in an evolutionary algorithm [26]. Drezewski and Siwik [18] designed a co-evolution multi-agent system, where an agent is a combination of the solutions and operations it can perform, such as clone, recombination and resources management. This approach was evaluated using ZDT benchmark, a multi-objective optimization benchmark.

Algorithms or Algorithm Components Agents: Some research treats other evolutionary components or even a full algorithm instance as agents. Zheng et al. [49] treat the whole population of solutions as an agent. This approach also keeps a master-agent is responsible for evolving the solutions for the original problem, and it is modeled as a top-level agent that is an asynchronous team, consisting of a population of complete solutions and a set of sub-agents working on the population. The authors evaluated their approach using mono-objective optimization problems such as Rosenbrock and Sphere. In [4] agents are the one which finds solutions for the optimization problem. There are two types of agents: master and slave. A master agent generates the population and divides them into sub-populations. It sends them to slave agents, responsible for executing conventional genetic algorithm steps on their sub-population, and periodically return their best partial results to the master agent. The master stores the partial results in lists. Denzinger and Offerman [17] present TECHS approach (TEams for Cooperative Heterogeneous Search), where Genetic Algorithms and Branch-and-Bound are considered as agents to solve the mono-objective job-shop-scheduling problem. Each agent gets the whole instance of the problem to solve, tries to solve the instance according to its search paradigm, and periodically interrupts its search in order to send information to other agents and to receive information from them. Chatzinikolaou and Roberteson [14] designed agents to contain a standard, canonical GA that acts on a local population of solutions, performing standard crossover and mutation on them. This paper also investigates the effectiveness with which reputation can replace direct fitness observation as the selection bias in an evolutionary multi-agent system. This is performed by implementing a peer-to-peer, self-adaptive genetic algorithm, in which agents act as individual GAs that, in turn, evolve dynamically themselves in real-time (namely the parameters employed by them). The evolution of the agents is implemented in two alternative ways: First, using the traditional approach of direct fitness observation (self-reported by each agent), and second, using a simple reputation model based on the collective past experiences of the agents. The authors validated their approach applying Rastringin, a mono-objective benchmark problem.

In [33], considered a genetic algorithm and a set of search techniques as agents to solve the mono-objective flexible job shop scheduling problem (FJSP). The genetic algorithm is employed for global exploration of the search space, the search set agents guide the research in promising regions of the search space and to improve the quality of the final population.

In [21] proposed a multi-agent hybrid algorithm composed by two agents: MOEA/D with a resource allocation mechanism (an evolutionary algorithm) and a local search method for the two-objective deteriorating scheduling.

Table 1 summaries all agent-based evolutionary algorithms, by classifying them according to how they implement agents in their design, if they implement these concepts on solutions, heuristics operator, algorithms instances. We can see that most papers employs solutions as agents and are designed for mono-objective optimization.

4.2   Multi-Agent Systems and Hyper-Heuristics

Several approaches have been proposed for combining MAS and Hyper-Heuristics. We can classify them as the following:

- Heuristic as agent: approaches which consider operators such as crossover and mutation as agents;
- Algorithm as agent: where a complete algorithm instance is considered as an agent;
- Both algorithms and heuristics as agents: approaches which consider as agents both different algorithm instances and heuristics as agents;
- Multi-objective: to identify which approach deals with multi-objective optimization;
- Cooperative or Competitive: if agents try to cooperate with each other or compete for computational resources.

Heuristic as Agent: Ouelhadj and Petrovic [36] employed search operators, such as Swap, Inversion, Insertion, and Permutation, as LLH agents to solve Permutation Flow Shop. This is a cooperative hyper-heuristic, where the heuristic agents perform a local search through the same solution space starting from the same or different initial solution and using different low-level heuristics. The agents exchanges their best solutions. After a generation, the best solutions are selected from all agents. This approach performs a greedy selection strategy to select an LLH to execute.

Meignan et al. [31] propose a selection hyper-heuristic where agents are responsible for concurrently explore the search space of an optimization problem in a cooperative way, where agents organized in a coalition cooperate by the exchanging of information about the search space and their experiences in order to improve agents behaviors. In order to generate new solutions, an agent uses several heuristics which are scheduled by an adaptive decision process, based on heuristic rules adapted along the optimization process by individual learning and. In this approach, a search agent keeps three solutions: the current, the best-found solution of the agent and the best solution of the entire coalition, and it can employ several operators on its current solution. This approach was applied to solve the Vehicle Routing Problem (VRP).

Table 1. Meta-heuristics papers classification

Paper	Solution as passive agent	Solution as active agent	Algorithms or components as agent	Multi-objective	Specially organized
Eiben et al. [19]	✓				
Socha and Kisiel-Dorohinicki [39]	✓				
Yan et al. [47]	✓				
Chira et al. [15]	✓			✓	✓
Ullah et al. [43]	✓				
Chalupa [13]	✓				
Wang and Wang [44]	✓				✓
Belkhelladi et al. [6]	✓				✓
Sun and Zhou [40]	✓				✓
Pan and Chen [37]	✓				✓
Zeng et al. [48]	✓				✓
Meignan et al. [30]		✓			
Jiang et al. [24]		✓		✓	
Huang et al. [23]		✓			
Corrêa et al. [26]		✓		✓	
Drezewski and Siwik [18]		✓		✓	
Zheng et al. [49]			✓		
Balid and Minz [4]			✓		
Denzinger and Offerman [17]			✓		
Chatzinikolaou and Robertson [14]			✓		
Nouri et al. [33]			✓		
Fu et al. [21]			✓	✓	

Algorithm as Agent: Cadenas et al. [9] introduced a cooperative multi-agent meta-heuristic approach, where agents communicate their best solutions using a common blackboard. This blackboard is monitored by a coordinator agent who is responsible to modify meta-heuristic agents behaviors based on fuzzy rules which take in account algorithms performance in the search. The authors tested their approach using 0/1 knapsack problems.

Malek [28] proposes a multi-agent hyper-heuristic to solve several combinatorial problems by considering GA, TS, SA, PSO, ACO as algorithm agents. In this approach, there is also a Problem agent, a Solution Pool agent (responsible to keep all solutions), and Adviser agent, an agent who provides parameter settings for the algorithms and receives reports from them. All algorithm agents of the same kind are associated with a common Adviser agent.

de Carvalho and Sichman [10,12] propose a Multi-Objective Agent-Based Hyper-Heuristic (MOABHH), an agent-based multi-objective hyper-heuristic focused on selecting the most suitable multi-objective evolutionary algorithm during execution time. MOABHH used the concept of voting to define which algorithm should receive a bigger participation in the generation of solutions. As a voting procedure, they applied the Copeland voting method and employed a set of voter agents responsible for evaluating algorithms (composed by NSGA-II, IBEA, and SPEA2) performance according to different quality indicators (composed by Hypervolume, Spread, RNI, GD, and IGD); these are usually used by the MOP community to compare the performance of MOEAs. After the voting, the most voted candidate received a bigger participation on the next offspring generation. In [11], the authors extended their work to solve four real-world engineering optimization problems. In this work, IGD and GD were replaced by HR and ER due to the fact of these indicators need previous problem knowledge, to make the approach applicable to real-world problems. This paper also set GDE3 as MOEA agent.

Acan and Lotfi [1] propose a collaborative hyper-heuristic architecture designed for multi-objective real-parameter optimization problems. In their approach, the population of solutions is split into sub-populations based on Pareto dominance, and then these sub-populations are assigned each one to a meta-heuristic agent, based on a cyclic or round-robin order, making meta-heuristics agents in this approach each operates on a sub-population in subsequent sessions. Meta-heuristic agents have their own population of non-dominated solutions extracted in a session, while there is also a global population of solutions keeping all non-dominated solutions found in the search. This study set MOGA, NSGAII, SPEA2, MODE, IMOPSO, AMOSA as meta-heuristic agents, and it was evaluated considering the CEC2009 benchmark.

Nugraheni and Abednego [34] propose an approach to select one of three agent Hyper-heuristics based on Genetic Programming (GPHH agent), Genetic Algorithm Hyper-Heuristic (GAHH agent), and Simulated Annealing Hyper-Heuristics (SAHH agent). These HH agents choose some low-level heuristics and work in search space of heuristics rather than a space of solutions directly.

Both Algorithms and Heuristics as Agents: Talukdar et al. [42] propose A-Teams, a synergistic team of problem-solving methods which cooperates by sharing a population of candidate solutions. In this approach, there is no coordination or planning mechanism, solutions are shared, through the central memory mechanism, allowing other agents to use these solutions in order to guide the search through promising search space, thus reducing the chances of being stuck at a local optimum. Aydin and Fogarty [3] extended A-Teams approach for solving Job Shop Scheduling. They employed as problem solving agents: SA (Simulated Annealing), TS (Taboo Search), HC (Hill Climbing), CSA (Simulated Annealing), CTS (Taboo Search), CHC (Hill Climbing), CHC2 (Hill Climbing), GA (Genetic Algorithm), NT (Improved version of CTS), and Damage. Barbucha [5] also extended the A-Team approach in order to create an Agent-Based Cooperative Population Learning Algorithm for the Vehicle Routing Problem with Time Windows. In his approach, the search is treated into stages, and different search procedures are used at each stage. The first stage is organized as an A-Team, where agents are used for improving the individuals stored in the common memory. In the second stage, the individuals in the population (common memory) are divided into subpopulations and allocated to a different set of A-Teams. In this level, each A-Team uses the same heuristics working under the same cooperation scheme. In the third stage, the sub-populations and the team of A-Teams architecture are also being employed. However, the process of communication among the set of A-Teams is used. The author evaluated his approach setting five problem-specific heuristics in the first stage, and a set of four Tabu Search and simulated annealing in the higher level.

Milano and Roli [32] presented the Multi-agent Meta-heuristic Architecture (MAGMA), a four-level architecture, with one or more agents at each level, where each level one or more agents act. The first level contains solution builders agents, responsible for providing feasible solutions for upper levels. The second level contains solution improvers, responsible for providing local search and solution improvements until a termination condition is verified. The third-level agents have a global view of the search space, or, at least, their task is to guide the search towards promising regions trying to avoid local optima. In the last level (Level-3) higher level strategies are described, such as a cooperative search and any other combination of meta-heuristics. The authors showed the three first levels are enough to describe standalone meta-heuristics and then evolutionary algorithms. Besides that, Level-3 can model coordinated cooperative hybrid meta-heuristics.

Talbi and Bachelet [41] propose a hybrid approach to solve the quadratic assignment problem by applying Tabu Search, Genetic Algorithm e KO (kick operator) as cooperative agents. The three heuristic agents run simultaneously and exchange information via an adaptive memory (AM). Each algorithm has a role: the Tabu Search is used as the main search algorithm, the Genetic Algorithm is in charge of the diversification and the Kick Operator is applied to intensify the search.

Martin et al. [29] propose a multi-agent hyper-heuristic where each agent implements a different meta-heuristic/local search combination. These agents also adapt itself along the search by using a proposed cooperation protocol based on reinforcement learning and pattern matching. Two kinds of agents are employed: launcher and meta-heuristic agents. The launcher is responsible for instantiating and keep the optimization problem, set up algorithms, and create initial solutions. The meta-heuristic agent contains an algorithm responsible for searching collectively for good quality solutions. The authors evaluated their approach using the mono-objective problems: Permutation Flow Shop, CVRP e Nurse Rostering and employing RandCWS, RandNEH as meta-heuristic agents.

Table 2 summaries all agent-based hyper-heuristics, by classifying them according to how they implement agents in their design. These paper are classified according to how they employ agents, if they are designed for multi-objective optimization, and if they are cooperative and competitive. Most of the research deals with mono-objective optimization and are cooperative.

Table 2. Hyper-heuristics papers classification

Paper	Heuristic as agent	Algorithm as agent	Algorithm and heuristic as agent	Multi-objective	Cooperative	Competitive
Talukdar et al. [42]			✓		✓	
Aydin and Fogarty [3]			✓		✓	
Barbucha [5]			✓		✓	
Milano and Roli [32]			✓		✓	
Talbi and Bachelet [41]			✓		✓	
Cadenas et al. [9]		✓			✓	
Malek [28]		✓			✓	
Ouelhadj and Petrovic [36]	✓				✓	✓
Meignan et al. [31]	✓				✓	
Martin et al. [29]			✓		✓	
de Carvalho and Sichman [10,11]		✓		✓	✓	✓
Acan and Lotfi [1]		✓		✓	✓	
Nugraheni and Abednego [34]		✓			✓	

5   Conclusions

This paper presented the synergy between Evolutionary Computation (EC) and Multi-Agent Systems (MAS) when both are used together to enhance the process of solving optimization problems.

We have proposed a criteria to characterize how agent theory is considered in order to design evolutionary algorithms, hybrid algorithms and hyper-heuristics by identifying which component is considered an agent and how this agent acts.

One can notice that this synergy has shown several interesting results, especially concerning the case of hyper-heuristics, due to the fact these approaches combine different heuristics and algorithm in order to better explore the search space. This set of components can naturally be designed as an agent-based system.

Acknowledgements. This study was financed in part by the Coordenação de Aperfeiçoamento de Pessoal de Nível Superior - Brasil (CAPES) - Finance Code 001. Vinicius Renan de Carvalho was also supported by CNPq, Brazil, under grant agreement no. 140974/2016-4.

References

1. Acan, A., Lotfi, N.: A multiagent, dynamic rank-driven multi-deme architecture for real-valued multiobjective optimization. Artif. Intell. Rev. **48**(1), 1–29 (2017)
2. Atallah, M.J.: Algorithms and Theory of Computation Handbook. CRC Press, Boca Raton (1998)
3. Aydin, M.E., Fogarty, T.C.: Teams of autonomous agents for job-shop scheduling problems: an experimental study. J. Intell. Manuf. **15**(4), 455–462 (2004)
4. Balid, A., Minz, S.: Improving multi-agent evolutionary techniques with local search for job shop scheduling problem. In: 2008 IEEE/WIC/ACM International Conference on Web Intelligence and Intelligent Agent Technology, vol. 2, pp. 516–521, December 2008
5. Barbucha, D.: A cooperative population learning algorithm for vehicle routing problem with time windows. Neurocomputing **146**, 210–229 (2014). Bridging Machine learning and Evolutionary Computation (BMLEC) Computational Collective Intelligence
6. Belkhelladi, K., Chauvet, P., Schaal, A.: An agent framework with an efficient information exchange model for distributed genetic algorithms. In: 2008 IEEE Congress on Evolutionary Computation (IEEE World Congress on Computational Intelligence), pp. 848–853, June 2008
7. Boussaid, I., Lepagnot, J., Siarry, P.: A survey on optimization metaheuristics. Inf. Sci. **237**, 82–117 (2013)
8. Burke, E.K., et al.: Hyper-heuristics: a survey of the state of the art. J. Oper. Res. Soc. **64**(12), 1695–1724 (2013)
9. Cadenas, J.M., Garrido, M.C., Munoz, E.: A cooperative system of metaheuristics. In: 7th International Conference on Hybrid Intelligent Systems (HIS 2007), pp. 120–125, September 2007
10. de Carvalho, V.R., Sichman, J.S.: Applying copeland voting to design an agent-based hyper-heuristic. In: Proceedings of the 16th Conference on Autonomous Agents and MultiAgent Systems, pp. 972–980 (2017)

11. de Carvalho, V.R., Sichman, J.S.: Solving real-world multi-objective engineering optimization problems with an Election-Based Hyper-Heuristic. In: International Workshop on Optimisation in Multi-agent Systems (OPTMAS 2018) (2018)
12. de Carvalho, V.R., Sichman, J.S.: Multi-agent election-based hyper-heuristics. In: Proceedings of the 27th International Joint Conference on Artificial Intelligence, pp. 5779–5780 (2018)
13. Chalupa, D.: Adaptation of a multiagent evolutionary algorithm to NK landscapes. In: Proceedings of the 15th Annual Conference Companion on Genetic and Evolutionary Computation, GECCO 2013, Companion, pp. 1391–1398. ACM, New York (2013)
14. Chatzinikolaou, N., Robertson, D.: The use of reputation as noise-resistant selection bias in a co-evolutionary multi-agent system. In: Proceedings of the 14th Annual Conference on Genetic and Evolutionary Computation, GECCO 2012, pp. 983–990. ACM, New York (2012)
15. Chira, C., Gog, A., Dumitrescu, D.: Exploring population geometry and multi-agent systems: a new approach to developing evolutionary techniques. In: Proceedings of the 10th Annual Conference Companion on Genetic and Evolutionary Computation, GECCO 2008, pp. 1953–1960. ACM, New York (2008)
16. Coello, C.: Evolutionary Algorithms for Solving Multi-objective Problems. Springer, New York (2007). https://doi.org/10.1007/978-0-387-36797-2
17. Denzinger, J., Offermann, T.: On cooperation between evolutionary algorithms and other search paradigms. In: Proceedings of the 1999 Congress on Evolutionary Computation-CEC99 (Cat. No. 99TH8406), vol. 3, p. 2324 (1999)
18. Drezewski, R., Siwik, L.: Agent-based multi-objective evolutionary algorithm with sexual selection. In: 2008 IEEE Congress on Evolutionary Computation (IEEE World Congress on Computational Intelligence), pp. 3679–3684, June 2008
19. Eiben, E.A., Schoenauer, M., Laredo, J.L.J., Castillo, P.A., Mora, A.M., Merelo, J.J.: Exploring selection mechanisms for an agent-based distributed evolutionary algorithm. In: Proceedings of the 9th Annual Conference Companion on Genetic and Evolutionary Computation, GECCO 2007, pp. 2801–2808. ACM, New York (2007)
20. Feigenbaum, E.A., Feldman, J., et al.: Computers and Thought. ACM, New York (1963)
21. Fu, Y., Wang, H., Tian, G., Li, Z., Hu, H.: Two-agent stochastic flow shop deteriorating scheduling via a hybrid multi-objective evolutionary algorithm. J. Intell. Manuf., 1–16 (2018)
22. Goldberg, D.E., Holland, J.H.: Genetic algorithms and machine learning. Mach. Learn. 3(2), 95–99 (1988)
23. Huang, J., Liu, J., Yao, X.: A multi-agent evolutionary algorithm for software module clustering problems. Soft Comput. 21(12), 3415–3428 (2017)
24. Jiang, S., Zhang, J., Ong, Y.S.: A multiagent evolutionary framework based on trust for multiobjective optimization. In: Proceedings of the 11th International Conference on Autonomous Agents and Multiagent Systems, AAMAS 2012, vol. 1, pp. 299–306, International Foundation for Autonomous Agents and Multiagent Systems, Richland, SC (2012)
25. Koza, J.R.: Evolution of subsumption using genetic programming. In: Proceedings of the First European Conference on Artificial Life, pp. 110–119 (1992)
26. de Lima Corrêa, L., Inostroza-Ponta, M., Dorn, M.: An evolutionary multi-agent algorithm to explore the high degree of selectivity in three-dimensional protein structures. In: 2017 IEEE Congress on Evolutionary Computation (CEC), pp. 1111–1118, June 2017

27. Maashi, M., Özcan, E., Kendall, G.: A multi-objective hyper-heuristic based on choice function. Expert Syst. Appl. **41**(9), 4475–4493 (2014)
28. Malek, R.: An agent-based hyper-heuristic approach to combinatorial optimization problems. In: 2010 IEEE International Conference on Intelligent Computing and Intelligent Systems, vol. 3, pp. 428–434, October 2010
29. Martin, S., Ouelhadj, D., Beullens, P., Ozcan, E., Juan, A.A., Burke, E.K.: A multi-agent based cooperative approach to scheduling and routing. Eur. J. Oper. Res. **254**(1), 169–178 (2016)
30. Meignan, D., Créput, J.C., Koukam, A.: A cooperative and self-adaptive meta-heuristic for the facility location problem. In: Proceedings of the 11th Annual Conference on Genetic and Evolutionary Computation, GECCO 2009, pp. 317–324. ACM, New York (2009)
31. Meignan, D., Koukam, A., Créput, J.C.: Coalition-based metaheuristic: a self-adaptive metaheuristic using reinforcement learning and mimetism. J. Heuristics **16**(6), 859–879 (2010)
32. Milano, M., Roli, A.: MAGMA: a multiagent architecture for metaheuristics. IEEE Trans. Syst. Man Cybern. Part B (Cybern.) **34**(2), 925–941 (2004)
33. Nouri, H.E., Belkahla Driss, O., Ghédira, K.: Metaheuristics based on clustering in a holonic multiagent model for the flexible job shop problem. In: Proceedings of the Companion Publication of the 2015 Annual Conference on Genetic and Evolutionary Computation, GECCO Companion 2015, pp. 997–1004. ACM, New York (2015)
34. Nugraheni, C.E., Abednego, L.: Multi agent hyper-heuristics based framework for production scheduling problem. In: 2016 International Conference on Informatics and Computing (ICIC), pp. 309–313, October 2016
35. Osman, I.H., Laporte, G.: Metaheuristics: a bibliography. Ann. Oper. Res. **63**, 511–623 (1996)
36. Ouelhadj, D., Petrovic, S.: A cooperative hyper-heuristic search framework. J. Heuristics **16**(6), 835–857 (2010)
37. Pan, X., Chen, H.: A multi-agent social evolutionary algorithm for resource-constrained project scheduling. In: 2010 International Conference on Computational Intelligence and Security, pp. 209–213, December 2010
38. Pearl, J.: Heuristics: Intelligent Search Strategies for Computer Problem Solving. Addison-Wesley Longman Publishing Co. Inc., Boston (1984)
39. Socha, K., Kisiel-Dorohinicki, M.: Agent-based evolutionary multiobjective optimisation. In: Proceedings of the 2002 Congress on Evolutionary Computation, CEC 2002, vol. 1, pp. 109–114, May 2002
40. Sun, H., Zhou, C.: Context-aware multi-agent model of microgrid optimization using fuzzy preferences evolutionary algorithm. In: 2013 Fourth International Conference on Intelligent Control and Information Processing (ICICIP), pp. 803–808, June 2013
41. Talbi, E., Bachelet, V.: COSEARCH: a parallel cooperative metaheuristic. J. Math. Model. Algorithms **5**(1), 5–22 (2006)
42. Talukdar, S., Baerentzen, L., Gove, A., De Souza, P.: Asynchronous teams: cooperation schemes for autonomous agents. J. Heuristics **4**(4), 295–321 (1998)
43. Ullah, A.S.S.M.B., Sarker, R., Lokan, C.: An agent-based memetic algorithm (AMA) for nonlinear optimization with equality constraints. In: 2009 IEEE Congress on Evolutionary Computation, pp. 70–77, May 2009
44. Wang, S., Wang, L.: A knowledge-based multi-agent evolutionary algorithm for semiconductor final testing scheduling problem. Knowl. Based Syst. **84**, 1–9 (2015)

45. Wolpert, D.H., Macready, W.G.: No free lunch theorems for optimization. IEEE Trans. Evol. Comput. **1**(1), 67–82 (1997)
46. Wooldridge, M.: An Introduction to Multiagent Systems. Wiley, Chichester (2009)
47. Yan, Y., Wang, H., Wang, D., Yang, S., Wang, D.: A multi-agent based evolutionary algorithm in non-stationary environments. In: 2008 IEEE Congress on Evolutionary Computation (IEEE World Congress on Computational Intelligence), pp. 2967–2974, June 2008
48. Zeng, C., Gu, T., Zhong, Y., Cai, G.: A multi-agent evolutionary algorithm for connector-based assembly sequence planning. Proc. Eng. **15**, 3689–3693 (2011). cEIS 2011
49. Zheng, Y., Xu, X., Chen, S., Wang, W.: Distributed agent based cooperative differential evolution: a master-slave model. In: 2012 IEEE 2nd International Conference on Cloud Computing and Intelligence Systems, vol. 01, pp. 376–380, October 2012

A Practice Report on the Active Learning Using Business Game for the Teacher Training Students

Hikaru Uchida$^{(\boxtimes)}$ and Katsutoshi Yuasa

Aoyama Gakuin University, Tokyo, Japan
uchida.h@aim.aoyama.ac.jp

Abstract. This paper describes a practice of business game using ICT (Information and Communication Technology, hereafter in ICT) prepared for the teacher-training course student. Since the Japanese primary and the secondary education, the environment of ICT in the classroom is improving. Furthermore, the forthcoming educational guidelines from the ministry are insisting on the active learning with ICT; it is more important to learn the active learning using ICT than ever. From the questionnaire survey, we examined how the students can use ICT for learning and what kind of difficulties the students have. As a result, it was speculated that students feel anxiety and difficulty in becoming teachers who do such classes because students have no experience of receiving active learning using ICT. In this study, as an example of active learning using ICT, we aim to make students experience business games using computer agents and aim to think more deeply about the possibility of using ICT for learning. In this paper, we describe the possibility of learning effect given by the design of the game using the computer agents. And we report the practice of business game using strategy agent. In the future, some concrete methods with ICT are also required such as mounting facilitating agents and simulating player agents in the game. This is research in progress.

Keywords: Teacher training course · Active learning · Business game · Computer supported collaborative learning

1 Introduction

The MEXT (The Ministry of Education, Culture, Sports, Science, and Technology) aims to realize education using ICT by 2020 [1]. The environment for classroom with the electronic blackboard, the digital textbooks, and the Internet are prepared. This vision is not merely the education using ICT, but also the spread of new educational methodology that the students can learn more proactively. In other words, the vision of MEXT aims to realize active learning more effectively by ICT than ever. We supported teacher training course for students with simulated lessons using ICT. In this context, the term "simulated lessons" refers to the lessons which the students play the role of teachers and students in the teacher training course lessons. We demonstrated the lessons using ICT such as e-blackboards and digital textbooks. And in next lessons, they did simulated lessons using ICT. Their lessons were used ICT, however, most of

© Springer Nature Singapore Pte Ltd. 2019
F. Koch et al. (Eds.): GEAR 2018, CCIS 999, pp. 42–48, 2019.
https://doi.org/10.1007/978-981-13-6936-0_5

the lessons were not active learning. It is difficult particularly for teacher training students to design useful lessons using ICT and to design lessons based on active learning. They may feel it as anxiety and threat because they lack knowledge of the new material "ICT" and learning experience using ICT [2]. Furthermore, even if they know methods of active learning, they had not experienced active learning when they were students. To realize the educational vision using ICT, it is important for students to integrate technological knowledge, pedagogical knowledge and contents knowledge [3]. Our conventional efforts were focused on only the technological knowledge, but not integrated with teaching knowledge. To fit this issue with the policy, instead of transforming the integrated knowledge, they needed experiences to participate in active learning using ICT as a student. We thought that students could integrate technological knowledge with teaching knowledge on active learning by experiencing business games using computer agents.

The rest of the paper is organized as follows: Sect. 2 gives an overview of agent-assisted active learning. Section 3 discusses practical reports on active learning using business games and its effects. Section 4 describes the issues and prospects of practicing the active learning in the teacher training course.

2   Agent Supported Active Learning

Gaming simulation is known as a teaching method that enhances learners' proactivity. Learners play a game with strategy, the interaction among players create competition. Learners play game with strategy, and the interaction among players create competition. U-mart is one of agent-based gaming simulation to learn the fundamentals of economics [4]. In U-mart, a virtual market is formed which is a combination of a computer agent and a human agent. Agent decisions are reflected in the market in real time. In this game, there are two learnings; such as to develop agents using strategies and to make decisions by looking at other agents. The learning design in which computer agents participate in human agents is a method by which computer agents support active learning. Human agents' strategy is uncontrollable, while computer agents' algorithmic strategy is controllable. There is a possibility of systematically designing active learning.

In this research, business games were employed as a method of active learning using ICT. Business games are materials that learn typical business models through gaming simulation. Like in U-mart, in the business game computer agents also support active learning. We expected that teacher training students could easily experience active learning using ICT.

3   The Active Learning Using the Business Game

We performed two lessons; in the first lesson, we conducted a simulation lesson using ICT and in the second lesson conducted a business game. In this session, (1) we introduce students' impressions to our simulation lesson, and (2) we report the computer agent's strategies and gaming practice. Finally, (3) we discuss students' awareness of ICT and active learning.

In the first lesson, we explained to students how to use ICT materials and showed a short lesson using ICT. We made presentations on the possibility of designing a new lesson using e-blackboards and digital textbooks.

We made a questionnaire about the impression of this demonstration. They mentioned "efficiency" as a good point of classes using ICT. Decreasing the burden on teachers and shortening preparation time is the excellent use of ICT. Moreover, they also mentioned that students' learning outcomes could easily be compared and evaluated. This awareness is about technical and pedagogical integrated knowledge.

On the other hand, as a difficulty, they mentioned the lack of computer skills and anxiety about dealing with trouble such as network connection and operation of computers. Also, some students pointed out the harmfulness given by using computers such as Health hazard such as eye deterioration. Their anxieties were similar to those reported by Mumtaz [2].

In the second lesson, the students experienced a business game called "restaurant game." The restaurant game is developed by Yokohama Business Game (YBG) [5] and operates in the WWW environment. The learner becomes the manager of the restaurant and decides the material cost, advertisement cost, and selling price. The learner's decisions are reflected in the total number of visitors to the restaurant market. The learner's decision is reflected in the number of customers visiting the restaurant and the market demand. Since the model of this game is simple and typical, it is known as teaching material for business beginners. Although the students are not affiliated with the school of business, it was speculated that they had the necessary knowledge to experience this game.

The purpose of learning using this game is not to learn the business model, but to think using the acquired knowledge. They were expected to notice that such thought learning is not natural.

One team consisted of 2 or 3 students and was divided into five teams together. We also mixed three computer agents. We also mixed three computer agents. Agent 1 made decisions without changing the decision to the end with default values. The default values are the market price written in the scenario to read before you start the game. Agent 2 copied the decision of the team having the most operating profit in the previous round as it is. The agent 3 decided on the price of the deviation value 60 from the selling price of the previous round. The agent assumed that the number of visitors is the average value of the previous round and decided the value adjusted so that the operating profit becomes zero. This was an agent that reproduced the high price strategy. However, it did not aim to obtain higher profits than the student team. We did not inform the students of the strategies of these agents and explained after the game.

The students used a calculation worksheet and a calculator application to think their team's decision. We did not in advance inform the learner of the end of the game. The game was ended when it was executed up to 7 rounds. Every team's decisions and outcomes were fed back each time the round was over. Three values for each team in all rounds were decided and three outcomes, sales, operating profit and cumulative operating profit, were indicated in Figs. 1 and 2. After presenting the graph on the e-blackboard to the whole class, its picture was distributed from the e-blackboard to the students' tablet and shared. In this game, they can calculate all outcome of each team during the game. However, sharing the graphs can give an overview of the decisions and outcomes of other teams in all rounds, and based on that, they can promote reflection of their team's decisions and strategies.

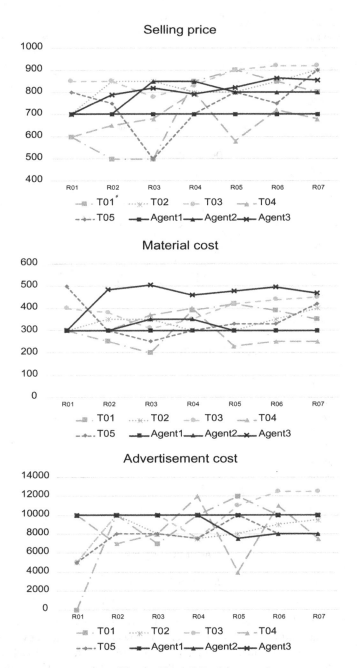

Fig. 1. Players' Decision

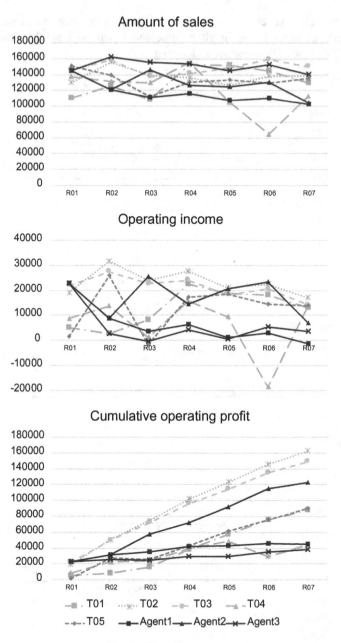

Fig. 2. Players' outcomes

The teams did reflection as looking at the graphs delivered to their tablet. They argued about what kind of strategy they were, and what kind of strategy was better. They sent their tablet screen that wrote the points of reflection to the e-blackboard. As shown in Fig. 3, answers of all teams were displayed on the e-blackboard. Each team presented their strategy, and according to their reflections, the class goal was considered to have been achieved. Team 2 presented that the point of their strategy is to observe the decisions of other players and the stable operation brought the cumulative operating profit at the final round. However, the operating profit in each round has slowed down. The operation of Team 1 that the selling price changed from a low price to a high price, and the margin increased slightly. Some difficult decision for the player, particularly beginners, in this business game is advertisement cost. Team 2 decided 0 (yen) in the first round and team 4 reflected that they did not think about advertisement cost adequately. They thought and decided in their own way, however, there was no awareness as regards computer agents. Our next challenge is to design computer agents to promote their thinking.

In their comments to active learning using ICT, some students pointed out educational efficiency such as quick responses and ease of consideration based on data, and other students were aware of learners' proactivity by gaming.

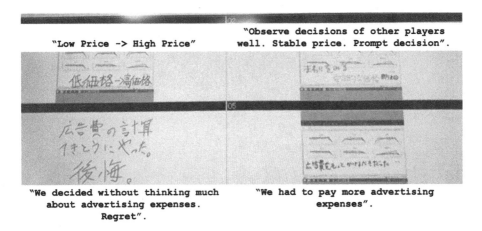

"Low Price -> High Price"

"Observe decisions of other players well. Stable price. Prompt decision".

"We decided without thinking much about advertising expenses. Regret".

"We had to pay more advertising expenses".

Fig. 3. Students' reflection and decisions on e-Blackboard

4   Conclusions and Future Work

In this paper, we treated a business game using agents as a method of "active learning using ICT" and reported practical practice for teacher training course students. According to their report, students were able to notice the goodness of gaming as an educational method and the possibility of utilizing ICT for designing learner-centered lessons.

The development of computer agents that promote learning challenges in the field of educational gaming simulation [6]. The future work is to design materials and lessons for teacher training students to be aware of the possibility of development of gaming simulation. We will build a teacher training program through gaming simulation and contribute to improving the knowledge and skills to design active learning using ICT.

References

1. MEXT: The Vision for ICT in Education - Toward the Creation of a Learning System and Schools Suitable for the 21st Century (2011). http://www.mext.go.jp/component/a_menu/education/micro_detail/__icsFiles/afieldfile/2017/06/26/1305484_14_1.pdf
2. Mumtaz, S.: Factors affecting teachers' use of information and communications technology: a review of the literature. J. Inf. Technol. Teacher Educ. 9(3), 319–342 (2000)
3. Mishra, P., Koehler, M.J.: Technological pedagogical content knowledge: a framework for teacher knowledge. Teachers Coll. Rec. 108(6), 1017–1054 (2006)
4. Matsui, H., et al.: The U-mart project: New research and education program for market mechanism. In: Arai, K., Deguchi, H., Matsui, H. (eds.) Agent-Based Modeling Meets Gaming Simulation, pp. 23–30. Springer, Tokyo (2005). https://doi.org/10.1007/4-431-29427-9_3
5. Yokohama Business Game. http://ybg.ac.jp/. Accessed 03 July 2018
6. Shirai, H.: The Front Line of the Business Game, 30(4), 409–416 (2015). in Japanese

Framework of Evaluating Business Partner Recommendation Beyond Industry Types Toward Virtual Corporation

Taisei Mukai[✉][iD]

Department of Inteligence System Science,
Tokyo Institute of Technology, Tokyo, Japan
mukai.t.ad@m.titech.ac.jp

Abstract. We propose a framework to evaluate a recommendation of unknown partners in an inter-business market by an artificial intelligence (AI) and a simulation. The reason is that unknown partner recommendation by AI such as a data mining or machine learning is difficult to evaluate because it is not possible to know correct partners in the real world. The framework is a flow of (1) proposing a method of recommending unknown business partners, (2) installing the recommendation method as AI into a firm agent, and (3) evaluating a recommended business partner by comparing performance between a recommendation method (machine learning, etc.) and an agent-based modeling simulation (ABMs). Since this framework can handle experiments assuming future situations, managers in a firm are possible to consider and judge recommendation methods and recommended partners according to virtual market conditions.

Keywords: Inter-firm market · Partner recommendation ·
Agent simulation · Machine learning · Virtual corporation

1 Introduction

The objective of this research is to examine the framework for evaluating and judging the partner recommendation for inter-firm trade beyond an industry type toward the virtual corporation by comparing performance between a recommendation method and an agent modeling simulation method [11].

The background of this objective is here. The industry 4.0 [2] has become a hot topic. This is the concept that firms can adapt flexibly to market changes by autonomously connecting factories, machines or workpieces. These connections will be eventually links between firms. The autonomous connection between firms was a concept that became famous as a virtual corporation [5] in the 1990's.

The virtual corporation is a type of firm in which a firm connects by using a network to create a virtual state like one firm in order to achieve its own

© Springer Nature Singapore Pte Ltd. 2019
F. Koch et al. (Eds.): GEAR 2018, CCIS 999, pp. 49–56, 2019.
https://doi.org/10.1007/978-981-13-6936-0_6

purpose. Even if a market environment changes, firms can adapt to the change by updating their purpose and their form of the virtual corporation [5]. A part of the virtual corporation is realized by consolidation of outsourcing such as a fabless company which do not have actual factories [3], the original equipment manufacturing (OEM) and the original design manufacturing (ODM) [21].

The reasons for such excitement of cooperation among firms are a management of resources via the Internet and the flexibility of business protocols such as the Web-EDI (Electronic Data Interchange) [19]. Furthermore, a recommendation of business partners by AI is also improved remarkably. Then this paper considers how the proposed framework with the AI judges recommended firms and supports the virtual corporation as under such recent technical conditions. This framework is used by the firm's managers to look for clients and providers and consider a trade in the various trade conditions. The case is as follows: (1) Managers look for a better business partner than the current business partner; (2) Managers restore business partners as soon as possible such as after an event of a disaster or look for providers that have high productivity after getting a big client; (3) Managers evaluate the recommended partner by changing a time step.

2   Related Study

We describe researches on business partner recommendation and a framework that can evaluate recommended partners. There are two cases of research on business partner recommendation as follows: (1) a method using information on a firm alone [8,9]; (2) a method using information on multiple firms [13]. The former is a method that learns a firm's performance such as firm credit or expert's estimation, with the information of a firm as features and applies it to the recommendation. The recommendation can be made by evaluating partner candidates prepared in advance. The latter method learns a label such as trade on or off of a trading pair (Client and Provider) by machine learning with binary discrimination using the firm information of the pair as features and recommends business partners by applying it. The paper focuses on the latter method.

However, most of the previous researches on the business partner recommendation are focused on acquiring firm information effective for selecting correct partners rather than recommending business partners. This is because machine learning and data analysis cannot easily judge whether the recommended unknown business partners are correct or not. These methods only learn past actual trades, then it is impossible to know positive labels of unknown trade. The above existing research also did not evaluate unknown partner recommendations.

Therefore, this research considers not only evaluating candidate firms directly using data, but also evaluating the performance of a business partner recommendation using the simulation framework of inter-firm trade in multi-perspective. The agent-based modeling simulation can express changes in the trading environment (requirement and demand) and can experiment with recommended partners by changing the conditions.

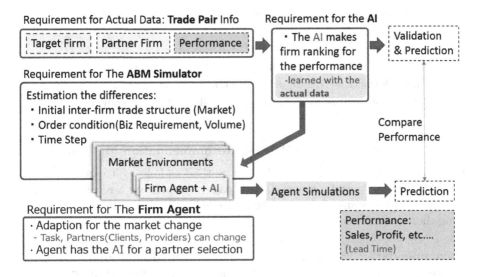

Fig. 1. The proposed framework

3 Proposed Framework

The proposed framework is for evaluating and judging whether the recommended new partners by the AI is good for improving the performance of a target firm. That is, the framework requirement is a comparison of the evaluation of unknown business partners by the AI and simulations in multi-perspective.

Briefly, first, prepare accounting data of the target firm and its business partner firms. Secondly, The AI learns performance indicators such as adjusted profit or sales of the target firm from the accounting data of this trade pair and evaluate it using prediction. Lastly, predict performance indicator by the ABMs in a market environment that firm agents equipped with the AI consist. It is a framework to judge business partners by trying various simulation conditions and comparing these multifaceted results with the results of machine learning (Fig. 1). Below we show the part of the requirements and examples of the proposed framework and how multiple evaluations are possible.

3.1 Requirements for Actual Data

The accounting data of the trading pair should have elements as features in Table 1 at least. Therefore, the framework requires three years' worth of accounting data of the target firm and its business partner firms to catch the sequential change of a state of the trade pairs. The procurement item types are a unique point in the elements.

Table 1. Firm features of a trade pair (Explanatory variable).

Attribute	Feature vectors
Categorical data	Own business item type, procurement item types
Firm information	Sales, profit, capital, worker number, ...
Composite variable	Growth rate, profitability, efficiency, stability, ...

3.2 Requirements for the Partner Recommendation Method (AI)

The recommendation method can be any method as long as it can calculate a ranking for a firm's performance. In this paper, we propose a recommendation algorithm with devising feature vectors (Table 1) and with a machine learning using these feature vectors (Fig. 1).

Recommendation Method Using a Firm Vector Model. First, express the features (Table 1) of a firm which deals with a target business item into feature vectors, and search and list firms of business partners based on the similarity (such as a cosin similarity, etc.) of the feature vectors of previous partners or desired business partners. As an unique idea, vectorize with procurement items of a firm and firm accounting figures (Table 1). As a result, it is possible to recommend firms that have a similar business output beyond its industry type and possibly the same business on its business scale. Firms are easy to trade with firms close to their size [1].

Recommendation Method Using a Machine Learning. For recommendation with a machine learning, it is recommended that both firm information of the new trade pair without trading in the past on real data for the two years after latest are used as feature vectors. The regression as the AI learns the ratio of sales or profit as a target variable (ex. (profit3 - profit2)/profit2 ...in Fig. 2), and after that, this regression predictor makes a ranking for unknown recommending partners in order of the performance (Fig. 2). The previous research [13] used trading pairs, but it was not adaptive to environmental changes because it learned past continued trade, and not leaned time-series trade for target variable and feature variables (vectors).

For example of a machine learning algorithm, a supervised machine learning is preferred. For example, a regression analysis: Multiple regression [4]; Support vector regression (SVR) [23]; or Random forest regression [10]; a multi-class classification: Multi-Class SVM [12], etc., or a artificial neural network can be learned by a regression or a multi-class classification [18,20] or its extended algorithm.

3.3 Requirements for the Firm Agent Model

Requirements for the agent model of an inter-firm is below: (1) Each firm agent adapts for the market change. The concept of a market change has elements of

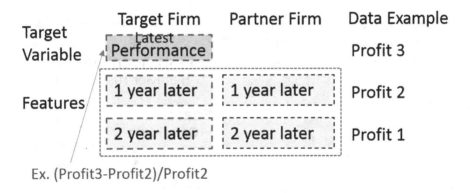

Fig. 2. Actual data of trade pair

business requirement change and business volume change. To meet the market change, the firm agent can change his business task item and his partners (clients and providers) to improve his performance indicators such as a profit or sales. The factors of the model are described in the previous researches [14,15]. This model Mukai proposed is made by expanding the division model that can search for business tasks based on the framework of enterprise behavior of previous research [17]. That's why we use this model.; (2) Each firm agent can equip the AI for the partner selection. See in Fig. 1. ABMs is easy to equip the AI.

Managers can use business partner recommendations for each business category in actual trade data. That is the business category level is a minimum firm agent size. Therefore, The abstraction level of the agent is the middle range [6].

3.4   Requirements for the Simulator

Requirements for the simulator is to estimate the difference below: (1) Difference of initial inter-firm trade structure; (2) Difference of order conditions (Business requirement and volume); and (3) Difference of time step; in this simulation. By comparing these simulation results with the results of machine learning, we judge whether the recommended firm is better or not (Fig. 1).

The simulator [14,15] satisfies these requirements and is characterized by an exploratory search of business items and partners to meets an environmental change in particular (Fig. 3).

Usage of the Framework for Evaluating a Partner Recommendation.
We describe a usage of the framework example: (1) Manager can compare simulation cases with and without dealing with recommended firms, in addition to evaluation of a machine learning; (2) Manager can compare simulation cases with and without dealing with business requirement change and volume change as the market change. This operation can express differences in business trade in growth areas and obscure areas. We can consider the differences of business

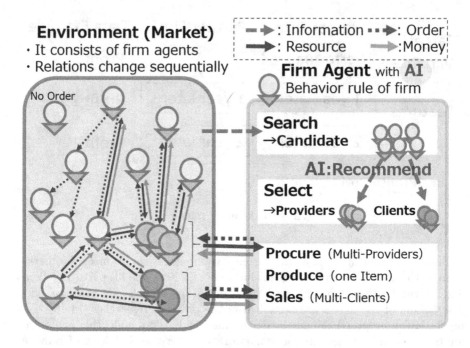

Fig. 3. Agent simulation model for inter-firm trade

partners in the future growth area or not; and (3) Even if the result of machine learning that learned most recent data is bad, the simulation case is a good in the long term or vice versa, or both are good or bad. You can evaluate the recommended partners from these differences.

4   Discussion and Future Work

We proposed the framework that can evaluate an unknown partner recommendation as an environmental adaptation. Even trade data that do not have a past trade can be evaluated according to virtual situations, so that decision can be supported diversely or autonomously in the future.

The objective of this framework is decision support for the manager in a firm, but we also consider to support an automatic trade between firms. In recent years, researches based on smart contracts with a blockchain such as the decentralized autonomous organization (DAO), the decentralized autonomous corporation (DAC) aiming to establish automatic trade are popular [16, 22]. Also, considering this autonomous trade as the virtual corporation, If the customer inputs what he wants in a system which deals with the virtual corporation, the trade path will be autonomously searched for, and the trade or service is established, executed or delivered as soon as conditions are matched. For example, the build to order (BTO) [7] realizes this partly so the system of the virtual corporation is easy to imagine.

In other words, the virtual corporation is autonomously constructed and accomplished by the customer just wishing for a service or what they want. The customer does not have to worry about how it was achieved. The proposed framework provides evidences to support this autonomous trade. Alternatively, it may be a role to stop a crazy autonomous trade.

Until now, ordinary customers were difficult to produce and buy what they wanted, unless they use joint purchasing. However, customers will be able to negotiate what they really want on the market through the virtual corporation system that the proposed framework supported.

References

1. Baum, J.A., Rowley, T.J., Shipilov, A.V., Chuang, Y.T.: Dancing with strangers: aspiration performance and the search for underwriting syndicate partners. Adm. Sci. Q. **50**(4), 536–575 (2005)
2. Brettel, M., Friederichsen, N., Keller, M., Rosenberg, M.: How virtualization, decentralization and network building change the manufacturing landscape: an industry 4.0 perspective. Int. J. Mech. Ind. Sci. Eng. **8**(1), 37–44 (2014)
3. Brown, C., Linden, G., Macher, J.T.: Offshoring in the semiconductor industry: a historical perspective. In: Brookings Trade Forum, pp. 279–333. JSTOR (2005)
4. Cohen, J.: Multiple regression as a general data-analytic system. Psychol. Bull. **70**(6p1), 426 (1968)
5. Davidow, W.H., Malone, M.S.: The virtual corporation: structuring and revitalising the corporation for the 21st century. New York (1992)
6. Gilbert, N.: Agent-Based Models, No. 153, Sage, London (2008)
7. Gunasekaran, A., Ngai, E.W.: Build-to-order supply chain management: a literature review and framework for development. J. Oper. Manag. **23**(5), 423–451 (2005)
8. Guo, X., Yuan, Z., Tian, B.: Supplier selection based on hierarchical potential support vector machine. Expert Syst. Appl. **36**(3), 6978–6985 (2009)
9. Guosheng, H., Guohong, Z.: Comparison on neural networks and support vector machines in suppliers' selection. J. Syst. Eng. Electron. **19**(2), 316–320 (2008)
10. Liaw, A., Wiener, M., et al.: Classification and regression by randomforest. R News **2**(3), 18–22 (2002)
11. Macal, C.M., North, M.J.: Tutorial on agent-based modeling and simulation. In: Simulation Conference, 2005 Proceedings of the Winter, 14 pp. IEEE (2005)
12. Mayoraz, E., Alpaydin, E.: Support vector machines for multi-class classification. In: Mira, J., Sánchez-Andrés, J.V. (eds.) IWANN 1999. LNCS, vol. 1607, pp. 833–842. Springer, Heidelberg (1999). https://doi.org/10.1007/BFb0100551
13. Mori, J., Kajikawa, Y., Kashima, H., Sakata, I.: Machine learning approach for finding business partners and building reciprocal relationships. Expert Syst. Appl. **39**(12), 10402–10407 (2012)
14. Mukai, T., Terano, T.: Modeling decentralized inter-organizational business structures through agent-based simulation. In: World Automation Congress (WAC), pp. 1–8. IEEE (2016)
15. Mukai, T., Terano, T.: Effects of trade environment in decentralized inter-organizational business structures through agent simulation. J. Adv. Comput. Intell. Intell. Inf. (JACIII) **22**(6), 933–942 (2018)

16. Norta, A.: Creation of smart-contracting collaborations for decentralized autonomous organizations. In: Matulevičius, R., Dumas, M. (eds.) BIR 2015. LNBIP, vol. 229, pp. 3–17. Springer, Cham (2015). https://doi.org/10.1007/978-3-319-21915-8_1

17. Okada, I., Ohta, T.: Psychological personality and organizational performance with MAS simulation. In: Agent-Based Approaches in Economic and Social Complex Systems, vol. 2, p. 35 (2002)

18. Ou, G., Murphey, Y.L.: Multi-class pattern classification using neural networks. Pattern Recogn. **40**(1), 4–18 (2007)

19. Ronchi, S.: The Internet and the Customer-Supplier Relationship. Routledge, New York (2018)

20. Specht, D.F.: A general regression neural network. IEEE Trans. Neural Networks **2**(6), 568–576 (1991)

21. Su, Y.f., Yang, C.: A structural equation model for analyzing the impact of ERP on SCM. Expert Syst. Appl. **37**(1), 456–469 (2010)

22. Swan, M.: Blockchain thinking: the brain as a DAC (decentralized autonomous organization). In: Texas Bitcoin Conference, Chicago, pp. 27–29 (2015)

23. Vapnik, V.: The Nature of Statistical Learning Theory. Jordan, M., Lauritzen, S.L., Lawless, J.L., Nair, V. (eds.) (1995)

Analysis of Researchers Using Network Centralities of Co-authorship from the Academic Literature Database

Masanori Fujita[✉]

Tokyo Institute of Technology, 4259-J2-12, Nagatsuta-cho, Modori-ku,
Yokohama, Kanagawa 226-8502, Japan
fujita.m.ai@m.titech.ac.jp

Abstract. Finding and encouraging young promising researchers is crucial to develop science and technology and to promote innovation. In this paper, I am to clarify requirements for researchers to conduct organizational Research and Development (R&D) and propose a quantitative method to evaluate researchers that satisfies the requirements to evaluate researchers in organizational R&D fields. A questionnaire survey was conducted to R&D institutions in life science and information technology fields to clarify the required competencies and careers of researchers for organizational R&D projects. The result of the survey suggests that the institution members require the researchers' competencies on not only "expertise of the research fields" but also "cooperativeness with others in the projects". Based on the result, I focus on network centralities of co-author networks and propose a new quantitative method to evaluate researchers by measuring the network centralities from the academic literature database.

Keywords: Academic literature database · Co-authorship · Network centrality

1 Introduction

To develop science and technology and promote innovation, organizational Research and Development (R&D), such as open innovation and industry-government-academia collaboration, has been becoming important, and finding and encouraging young promising researchers is a crucial issue. To evaluate such researchers, citation-based indexes such as "h-Index" are often used [1]. These are quantitative indexes showing impact of their research achievements, but difficult to evaluate young researchers who don't have sufficient research achievements, because the citation-based indexes are lagging indexes which follow the research achievements [2]. To complement this problem, peer reviews are also used. In the peer reviews, it is possible to evaluate young researchers' personal competencies, but difficult to conduct quantitative and widely-covering evaluation of many researchers without heavy burden imposed on evaluators.

With such a background, in this paper, in order to evaluate researchers in collaborative research, we clarify requirements for desired researchers, and propose a new

© Springer Nature Singapore Pte Ltd. 2019
F. Koch et al. (Eds.): GEAR 2018, CCIS 999, pp. 57–63, 2019.
https://doi.org/10.1007/978-981-13-6936-0_7

quantitative and widely-covering evaluation method of researchers that satisfies the requirements without heavy burden on the evaluators.

In the following sections, I describe methodology of the research of this paper. Next, I briefly show a result of our survey to clarify the requirements of the researchers, then propose a new quantitative evaluation index using network centralities constructed from the academic literature database. Finally, I illustrate two applications of the index.

2 Methodology

In this paper, we conduct a questionnaire survey to clarify requirements for promising researchers. As a result of the survey, it is found that "cooperativeness" is one of very important requirements for researchers. Based on the result, as an evaluation index of researchers, we focus on "betweenness centrality of co-author networks" from academic literature database which indicates cooperative competency of researchers in collaborative research.

Allen found star researchers, who are the core of communication among researchers, improve the performance of R&D, and called such star researchers as "Gatekeepers" [3]. Allen's Gatekeepers as illustrated in Fig. 1. Have continuous contact with researchers inside and outside the organization and are mediators who translate and transmit external information into the internal organization. On the other hand, Fig. 2 shows one example of co-author networks which are collaboration networks of researchers where researchers with high betweenness centralities are shown as dark colored spots. In comparison between Figs. 1 and 2, the researchers with high betweenness centralities are found to be equivalent to Gatekeepers who mediate between other researchers.

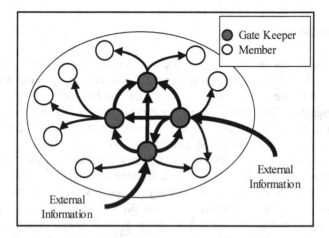

Fig. 1. Allen's "Gatekeepers" are star researchers who are the core of communication among researchers and improve the performance of R&D. Gatekeepers have continuous contact with researchers inside and outside the organization and translate and transmit external information into the internal organization (Allen 1973).

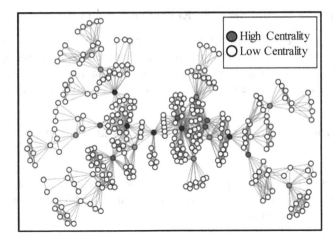

Fig. 2. One example of co-author networks which are collaboration networks of researchers. Researchers with high betweenness centralities are shown as dark colored spots and they are equivalent to Allen's "Gatekeepers."

In this paper, betweenness centralities of co-author networks are analyzed in about 4 million literatures of academic literature database (JSTPlus) provided by Japan Science and Technology Agency (JST) to evaluate researchers.

3   Requirements for Desired Researchers and Proposal of an Evaluation Index of Researchers

In this section, I briefly show a result of our survey to clarify the requirements of the researchers and then propose a new evaluation index using network centralities.

3.1   Requirements for Desired Researchers

To clarify the requirements for desired researchers in organizational R&D, we conducted a questionnaires survey [4]. In the questionnaire, items of the requirements for researchers are divided into their competencies and their careers. The competencies are subdivided into 3 categories of items; items related with research subjects, items related with other researchers and items related with themselves [5–7]. Target of the survey are members of "Life Intelligence Consortium" which is R&D consortium consisting of life-science institutions such as pharmaceutical companies and IT institutions such as electronics manufacturers. Number of the members is 526 and number of the respondents is 104. As a result of the survey, requirements for the researchers are found as follows; (1) competencies are more important than careers, (2) among the competencies, personal qualities of "cooperativeness" and "autonomy" are important in addition to "expert knowledge" and "professional skills."

3.2   Betweenness Centralities as an Evaluation Index of Researchers

Based on the result of the survey that "cooperativeness" is an important competency for researchers in collaborative research, we proposed "betweenness centrality of co-author networks" as a quantitative and widely-covering evaluation index of researchers which indicates cooperative competency of researchers in collaborative research.

4   Applications of Network Centralities as an Evaluation Index

In this section, I show two applications of the proposed evaluation index. One is "evaluation of researchers selected for funding programs", and the other one is "search of young promising researchers".

4.1   Evaluation of Researchers Selected for Funding Programs

We analyzed the centralities of co-author networks constructed from literatures in biology field of JSTPlus [8]. As the result, centralities of "Japan Society for the Promotion of Science (JSPS) Research Fellows", who are selected by JSPS as promising researchers, are found transiting significantly in logit model. In addition to the centralities of JSPS Research Fellow, as shown in Fig. 3, centralities of "PRESTO" researchers and "CREST" researchers, who are both selected as excellent researchers by JST, are also rapidly growing after the selection by each grant program, while they are showing different transitions from each other before the selection.

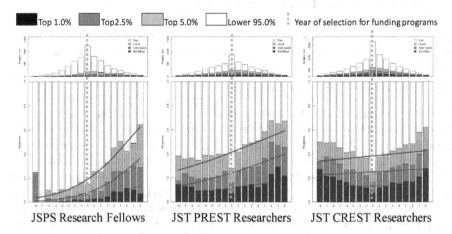

Fig. 3. Transition of network centralities of JSPS Research Fellows, PREST researchers and CREST researchers are rapidly growing after the selection by each grant program, while they are showing different transitions from each other before the selection.

4.2   Search for Young Promising Researchers

Figure 4 shows the result of comparative analysis between the centralities of JSPS Research Fellows and the centralities of all researchers appeared in biology and physics fields of JSTPlus. In both fields, the centralities of JSPS Research Fellows are found transiting in logit model and growing more rapidly than those of all researchers.

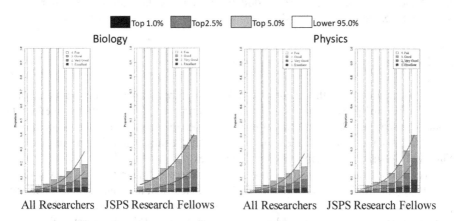

Fig. 4. Centralities of JSPS Research Fellows in biology and physics fields of JSTPlus are transiting in logit model and growing more rapidly than centralities of all researchers which are growing in linier model.

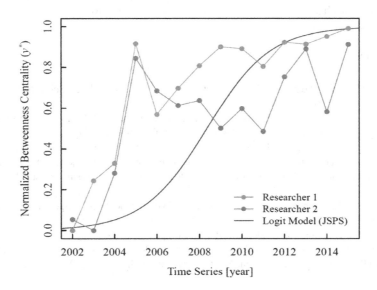

Fig. 5. Two examples of promising researchers extracted by the decorrelation test of the centralities of 4 years period from their first literatures publications with those of JSPS Research Fellows'

Based on the result, we propose a method to search promising researchers by using network centralities [9]. In this method, we extract researchers with transition of the centralities significantly similar to transition of those of JSPS Research Fellows by decorrelation test between the centralities of targeted researchers and the centralities of JSPS Research Fellows.

Figure 5 shows 2 examples of promising researchers extracted by the decorrelation test of the centralities of 4 years period from their first literatures publications with those of JSPS Research Fellows'.

5   Conclusion and Future Works

In this paper, I clarified the requirements for desired researchers in the organizational research by the questionnaire survey. Then I proposed the quantitative method to evaluate researchers by measuring transition of betweenness centralities of co-author networks as an evaluation index that satisfies the requirements for the researchers. Finally, I illustrated two applications of the index.

We focused on biology and physics fields where collaborative research style is usually common. Meantime, in interdisciplinary fields such as bioinformatics, not only collaborative research in a single field but also fusion research across multiple fields may be necessary or effective. Moreover, in the case of JSPS Research Fellow, the researchers are supposed to change research institutions after the selection, and these change of research institutions may create new research opportunities and affect the growth of the researchers. So, in the future, I plan to conduct research on causal relationship between the growth of researchers and distributions of their research fields they engage or changes of research institutions they belong to.

References

1. Hirsch, J.E.: Does the h index have predictive power? Proc. Natl. Acad. Sci. U.S.A. **104**(49), 19193–19198 (2007)
2. Zare, R.N.: Assessing academic researchers. Angew. Chem. Int. Ed. **51**, 7338–7339 (2012)
3. Allen, T.J.: Managing the Flow of Technology. MIT Press, Cambridge (1979)
4. Fujita, M., Ishido, K., Suzuki, Y., Inoue, H., Terano, T.: A questionnaire survey study to clarify desired characteristics of promising researchers for research projects in the life science domain. In: National Conference of JASMIN 2018 Spring at University of Tsukuba, Tokyo (2018)
5. Spencer, L.M., Spencer, S.M.: Competence at Work: Models for Superior Performance. Willy, New York (1993)
6. OECD: The Definition and Selection of Key Competencies (2003)
7. Matsushita K.: A New Framework for Competencies: 3-3-1 Model. Center for the Promotion of Excellence in Higher Education, Kyoto University, vol. 22, pp. 139–149 (2016)

8. Fujita, M., Inoue, H., Terano, T.: Evaluating funding programs through network centrality measures of co-author networks of technical papers. In: The 2nd International Workshop on Application of Big Data for Computational Social Science 2017, IEEE BigData 2017 Workshop, Boston (2017)
9. Fujita, M., Inoue, H., Terano, T.: Searching promising researchers through network centrality measures of co-author networks of technical papers. In: The 4th IEEE International COMPSAC Workshop on Social Services through Human and Artificial Agent Models (SSERV 2017), Torino (2017)

Debriefing Framework for Business Games Using Simulation Analysis

Takamasa Kikuchi[1][(✉)] ⓘ, Yuji Tanaka[2], Masaaki Kunigami[2] ⓘ,
Takashi Yamada[3], Hiroshi Takahashi[1], and Takao Terano[4] ⓘ

[1] Keio University, 4-1-1 Hiyoshi Kohoku-ku, Yokohama, Kanagawa, Japan
takamasa_kikuchi@keio.jp
[2] Tokyo Institute of Technology, 4259 Nagatsuta-cho, Midori-ku,
Yokohama, Kanagawa, Japan
[3] Yamaguchi University, 1677-1 Yoshida, Yamaguchi-shi, Yamaguchi, Japan
[4] Chiba University of Commerce, 1-3-1 Konodai, Ichikawa-shi, Chiba, Japan

Abstract. Researchers are aware of the importance of debriefing in gaming. However, there has been little progress in establishing a methodology for evaluating player behavior. Therefore, we propose a framework to support the evaluation of player behavior in business games using computers. Specifically, we introduce a simulation analysis methodology that involves the following steps: (1) constructing an agent-based model based on the subject of the business game and categorizing simulation-logs; and (2) mapping logs of player behavior onto typed results. In this way, the positioning of the player in the overall simulation scenario is visualized, and a range of possible results is presented. Both players and facilitators receive information that is useful for debriefing.

Keywords: Business game · Agent-based simulation · Simulation analysis

1 Introduction

Researchers are aware of the importance of facilitation and debriefing in relation to gaming [1]. However, facilitation and debriefing tend to be regarded as an "art" that is largely dependent on the experience and skill of the facilitator. In evaluating the behavior of a player, it is common to analyze the contents of the individual's play-log. However, the analysis of the entire game involves the following difficulties: (1) collection of large number of play-logs, (2) biased samples can possibly exist, (3) it is difficult to create a comprehensive list of all possible results. As a result, there are indications that the axis of evaluation is weighted towards results in terms of wins/losses [2]. Hence, there has been little progress in relation to the establishment of a methodology for analyzing and evaluating a player's judgment and behavior in gaming.

This study proposes a framework to properly evaluate subject behavior in computational business games (see Sect. 2.1). Specifically, we introduce a simulation analysis methodology that involves the following steps: (1) an agent-based simulation of/on the corresponding/underlying business game is pursued exhaustively and the simulation logs are then categorized, and (2) the results (or game logs) of laboratory

© Springer Nature Singapore Pte Ltd. 2019
F. Koch et al. (Eds.): GEAR 2018, CCIS 999, pp. 64–76, 2019.
https://doi.org/10.1007/978-981-13-6936-0_8

experiments are mapped on the categorized computational results so that the behavior of subjects is identified. These steps enable to present to the player and facilitator and visualize the position of the player with respect to these possible results.

Then, in this study, "Log" is defined as the state variables and the decision-making of human subjects and agents in a business game or an agent-based simulation. One log is outputted by playing a business game or executing an agent-based model (hereinafter "ABM").

The rest of this paper is organized as follows. Section 2 provides a summary of related research and definitions of key terms. Section 3 presents the proposed methodology. Section 4 provides an example of its application. Section 5 provides a summary and concludes.

2 Related Work and Definitions

2.1 Gaming

Gaming is a technique that originated in military training exercises. In recent years, gaming targeting the field of business has attracted increasing attention [3].

Normally, a professor serves as the facilitator of a game, while students (players) compete in accordance with the prescribed rules either as individuals or as members of a team to achieve the pre-determined objective of the game. This allows participants to learn by themselves. The process of learning through games consists of three steps: background description (briefing), execution (gaming), and reflection (debriefing) [3].

While in the beginning the style of gaming experiment was so-called "paper-and-pencil" experiment where cards, dices, boards and the related device were used, the computerized experiment has been widely used later.

In addition, virtual agents have been introduced to the gaming environment to improve the feasibility of game execution and to clarify the actions of participating parties [3]. Examples include Alexander Islands [4] and the U-Mart project [5].

However, a suitable debriefing method has not yet been established. In this study, we propose a support framework for debriefing following business games using computers that can be played not only by humans but also by virtual agents.

In this study, the result of a game played by a human is a "play-log," while the result of a game played by an agent is a "simulation-log." Furthermore, a platform on which humans and agents are able to coexist and participate in games is called a "gaming simulator."

2.2 Evaluation of Players in Business Games

The evaluation of player behavior in business games includes protocol analysis, understanding degree sheets, and approaches using performance sheets [2]. Then, performance sheets [2] are intended to grasp the reasons of decision-making and to visualize the behavior by letting the player answer the fixed questions in each round of the business game using computers. An example of the questions is as follows:

> *Question 1* "Input to the game":
>   How did you try to change the decision items?
> *Question 2* "Referenced indicators":
>   What are the management indicators that base the judgment of Question 1?
>   What was the status of the management index of Question 2?
> *Question 3* "Target indicators":
>   Which management indicators were tried to be changed by Question 1 after receiving Question 2?

This approach is an attempt to visualize the behavior and judgment of the player during the game.

In this study, we present an overview of the scenarios and simulation paths that business games can follow. Furthermore, we visualize the position of the player with respect to the range of possible results.

2.3   Gamification

Gamification is a methodology that considers social and organizational behaviors and activities as a game, interprets the world, and helps design new/better systems [6].

To develop gamification as a system, Terano and Koyama have proposed an approach such that a loop consisting of actual problem → gaming → agent-based simulation (ABS) → story/scenario preparation → grounding to reality (Fig. 1) [7].

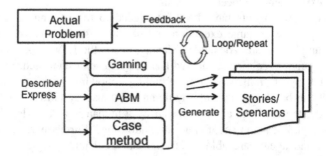

Fig. 1. Approach to realize the concept of gamification: a combination of gaming and ABM.

For example, the following studies [8–10] have (1) tried to combine business cases and gaming, (2) tried to match virtual cases with actual business cases generated by ABM simulation-logs, and (3) tried to use gaming to promote the understanding of ABMs.

In this study, we propose an evaluation framework for business games based on ABM simulation analysis.

3 Methodology

3.1 Outline

Figure 2 shows a summary of the methodology used in this study. The proposed method consists of the following six steps:

<Step 1>
Determine the theme of the business game.

<Step 2>
Create an ABM based on the theme of the business game. Generate numerous simulation-logs from play by agents in the virtual space of the model.

<Step 3>
Categorize the results of <Step 2> and grasp an overall picture of possible scenarios. In this study, we apply the clustering method detailed in Sect. 3.2. The result of typing is displayed as a log cluster. In addition, set/determine the criteria from the micro-level attributes of each agent and the emerged macro-level attributes in the simulation-log.

<Step 4>
Extend the ABM constructed in <Step 2> so that humans are able to substitute for some agents.

<Step 5>
Map human play-logs to log clusters (see Fig. 3). Specifically, plot the results of play-logs on the coordinate axes set when clustering is performed in <Step 3>.

<Step 6>
Perform an analysis for debriefing. The methodology for this step is shown in Sect. 3.3.

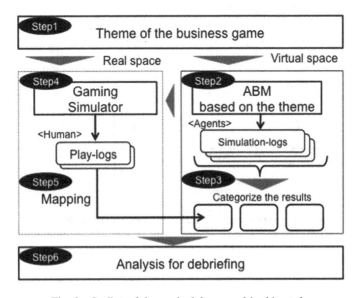

Fig. 2. Outline of the methodology used in this study.

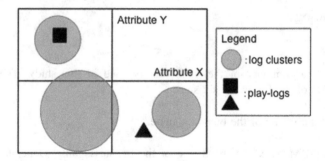

Fig. 3. Schematic diagram of the mapping of play-logs to log clusters. Obtain the overall scenario by clustering. Plot the results of individual player's play-logs (■ and ▲) for typing.

3.2 Simulation Analysis

To type scenario and simulation paths as described in <Step 3> above, this study uses the log cluster method [11].

In this method, we apply hierarchical clustering (the Ward method) to the simulation-logs of the agent model, which are then categorized (see Fig. 4). Then, the distance between clusters is calculated using the dynamic time warping method. The illustrated log cluster represents the frequency of occurrence not indicate the boundary between clusters.

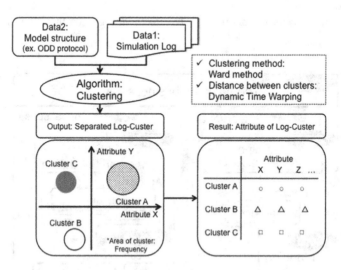

Fig. 4. Simulation analysis method adopted in this study [11].

3.3   Analysis for Debriefing

The analysis for the debriefing in <Step 6> is undertaken as follows:

(1) The typed scenario/simulation pass (log cluster) presented in Fig. 4 shows the possible scenarios in the business game and the overall simulation pass.
(2) To obtain a comparison between the play-log and the log cluster, the distance (similarity) between the play-log and each log cluster is calculated. In addition, possible results other than those achieved by human players are presented.
(3) Furthermore, the comparison between play-logs indicates whether there are any differences between the players, and if so, whether they can be explained by differences in the log clusters.

4   Demonstration

4.1   Business Game Theme

The subject of the business game used in this study is the influence of the investment behavior of financial institutions in relation to marketable assets on the stability of the financial system. The player assumes responsibility for asset and liability management (ALM) in a financial institution and makes decisions concerning the investment of securities.

Prior research in this field discusses trade-offs between maximizing the profit of financial institutions and the stability of the financial system. For example, research on the financial crisis after the Lehman shock has focused on (1) leveraging problems of financial institutions and (2) the impact of financial regulation [12].

4.2   ABM Based on the Game Theme

Figure 5 shows the ABM constructed on the basis of the business game theme used in this study. This is based on the ABM used in [13], which examines the chain reaction in terms of the failure of financial institutions.

4.2.1   Agents

Agents are core financial institutions that have a simplified balance sheet and basic financial indicators such as the capital adequacy ratio and ROE (Return On Equity). The balance sheet includes marketable and non-marketable assets, cash, and deposits. Based on their balance sheet, each financial institution is classified as either a fund surplus entity or a fund shortage entity.

4.2.2   Networks

Each financial institution borrows short-term funds through the interbank network. This mimics the role of the real-world money market (mainly the call market in Japan) [14] in adjusting for excesses or shortages of funds between fund surplus entities and fund shortage entities.

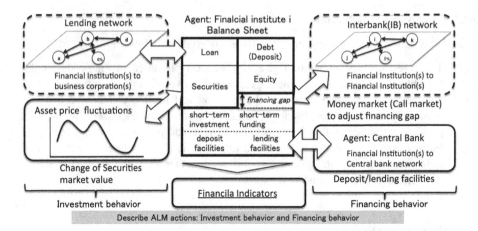

Fig. 5. ABM showing the chain reaction relating to the failure of financial institutions presented in a previous work [13]. The agents are core financial institutions, which borrow short-term funds through the interbank network. They make decisions relating to (1) investment behavior related to the sale and purchase of marketable assets, and (2) cash-flow behavior to address funding excesses and shortfalls. Their financial situations change in response to fluctuations in the prices of marketable assets, which represents the external environment, and in some cases, a chain collapse of financial institutions occurs. In this study, we focus on investment behavior.

4.2.3   Agent Decision-Making

Each financial institution makes decisions regarding (1) investment behavior (purchases and sales of marketable assets) and (2) financing behavior (lending/borrowing short-term funds). In this study, we focus on investment behavior based on the theme described in Sect. 4.1.

In the proposed model, financial institutions make investment decisions in relation to marketable assets on a step-by-step basis in accordance with their own market trends based on various management constraints: (a) capital adequacy ratio, (b) VaR, and (c) ROE/budget). In this study, after setting (a) and (b) as absolute constraints, decisions are made regarding the purchase, sale, and maintenance of marketable assets so as to maximize (c), taking into account the market view [13].

4.2.4   External Environment

Price fluctuations in marketable assets are given as the external environment. Each financial institution adjusts their financial indicators such as the capital adequacy ratio through regular revaluation of the marketable assets they hold.

4.2.5   Bankruptcy Mechanisms

In the proposed model, possible causes of bankruptcy in financial institutions are as follows: (1) excessive debt, (2) decrease in the capital adequacy ratio to below a certain level, and (3) lack of sufficient funding to continue procurement. The first item refers to general loans and interbank loans that are written off because they cannot be absorbed by the company's capital balance. The second item refers to a capital adequacy ratio consistent with uniform international standards, which currently exceeds 8% [15]. The third item refers to situations in which a company cannot obtain funding necessary to overcome a shortfall on the short-term money market.

4.2.6 Simulation Settings

Specific parameters are based on [13] and are targeted at 20 financial institutions that are complete graphs (see Table 1). In our simulation, we select an arbitrary financial institution and analyze the sensitivity of the results after changing the initial holding of marketable assets. The financial institution does not make any investment decisions during this step. This is consistent with the setting used in business games played by human players, as described in Sect. 4.5. The aim is to analyze the results with a view to understanding the various decisions that players can potentially make.

Table 1. The main parameters used in this study.

Parameters	Value
Number of institutions N	20
Interbank network $W^{Interbank}$	Complete graph
Cash CA	10–25, uniform distribution
Non-marketable asset nonMA	100, constant
Marketable asset MA	20–50, uniform distribution
Capital adequacy ratio CAR	12%–22%, uniform distribution

The price of risky assets is assumed to be represented by the following discretized stochastic differential equation [16]:

$$P(t,j) = P(t-1,j) + r_f P(t-1,j)\Delta t + \sigma P(t-1,j)\tilde{\varepsilon}\sqrt{\Delta t}$$

Where t is the period $(t = -m+1, \ldots, 0, 1, \ldots, T)$, j is the trial number, $P(t, j)$ is the price of the marketable asset (j times, step t) ($P(0) = 100$), r_f is the risk-free rate (%), σ is volatility (%), and $\tilde{\varepsilon} \sim N(0, 1)$. In this simulation, we set 1 step = 1 day = 1/250 year and $\Delta t = 1/250$, $T = 125$ (assuming six months is the budget-closing period for a bank account) and $m = 16$. Additionally, taking into account long-term government bond yields and the stock market in each country, $r_f = 2\%$ and $\sigma = 25\%$. We generate 100,000 sample paths, and the prices of marketable assets in the final step are used to select the time series with the lowest prices (Fig. 6).

4.3 Classification of the Results of Simulation-Logs

One thousand trials were conducted using the settings and parameters described in Sect. 4.2.6. We applied hierarchical clustering (the Ward method) as described in Sect. 3.2 to the simulation-log and categorized it into three clusters.

The results of the clustering are shown in Fig. 7. The vertical axis shows the initial holding of marketable assets by the financial institution that was selected, and the horizontal axis shows the number of steps of failure of the agent. The parameter set used

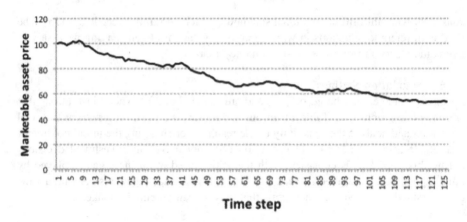

Time step

Fig. 6. Time series of marketable asset prices.

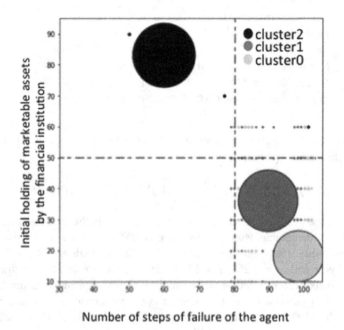

Number of steps of failure of the agent

Fig. 7. Clustering of the simulation-log.

in this simulation was classified into the following three log clusters according to the quantity of marketable assets and the number of failed steps. Each cluster can be roughly interpreted as follows:

cluster 0: The outstanding balance of marketable assets is restraint, and in the case of bankruptcy caused by burning out funds lent to other companies that failed to chain (hereinafter referred to as "Other responsible bankrupt case").

cluster 1: The outstanding balance of marketable assets is moderate, and in cases where collapse took place almost simultaneous with other companies (hereinafter "Simultaneous bankruptcy case").

cluster 2: In the case where the balance of marketable assets is large and the bankruptcy has become solely due to excessive leverage (hereinafter referred to as "Single collapse case").

4.4 Gaming Simulator

We extend the ABM that was proposed in Sect. 4.2 so that humans can replace agents in decision-making. The procedure regarding the gaming simulator created in this study is as follows:

4.4.1 Determining the Role of the Player
The human player is in charge of one of the 20 financial institutions listed in Sect. 4.2.6, while the other 19 companies are in the charge of an agent. Two students from the Tokyo Institute of Technology each play the game once.

4.4.2 Presentation of Initial Competitive Environment
We present the initial competitive environment to the players on paper, including the financial situations (such as the balance sheet structure) of their own financial institution and their competitors and operational restrictions (such as ROE/budget constraints).

4.4.3 Player's Investment Decisions
We determine the initial marketable asset balance owned by the financial institution in charge and entering it into ABM. For the sake of simplicity, only the marketable asset balance at the start step is the decision item. This is a simplification of the investment and budget plan at the beginning of the period in terms of ALM.

4.4.4 Start of the Simulation Process
The ABM simulation, which comprises 125 steps, is started. In this study, financial institutions controlled by humans only make an initial investment decision and do not make subsequent investment decisions.

4.5 Mapping Human Play-Logs

Using the gaming simulator presented in Sect. 4.4, an individual trial was conducted with two players (trial 1 and trial 2). Figure 8 shows the mapping of individual play-logs by humans onto the simulation results described in Sect. 4.3. The distance from the center of gravity of each log cluster is shown. It was found that trial 1 was close to *cluster 2* and trial 2 was close to *cluster 0*.

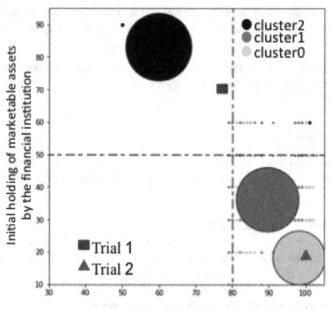

Number of steps of failure of the agent

	cluster0	cluster1	cluster2
Trial 1	3.47	3.63	2.12
Trial 2	1.69	2.36	2.83

Fig. 8. Results of mapping play-logs by humans.

4.6   Analysis for Debriefing

The analysis of this example is shown below:

(1)   Based on the results presented in Sect. 4.3, there were three possible results: other responsible bankrupt case, simultaneous bankruptcy case, or single collapse case.
(2)   At the same time, trial 1 (trial 2) presented other results that were "possible".
(3)   Furthermore, the difference between trial 1 and trial 2 is represented by the difference in the log cluster (failed case) to which it belongs, and was found to be the result of a difference in the initial amount of marketable assets.

In (1) above, we obtained an overall image of the scenario/simulation results without collecting numerous play-logs created by humans.

In addition, in (2) above, we examined the results of the human players to determine which of the results were similar to those of multiple trials involving agents.

Furthermore, in (3) above, we analyzed factors other than results in terms of wins and losses in relation to differences in player's results. As noted above, the proposed methodology can provide information that is useful for both the facilitator and the player at debriefing.

5   Concluding Remark

In this study, we propose a framework for evaluating the behavior of players in business games and find that (1) it is possible to present the whole scenario simulated path in the business game and the whole image of the simulation pass, (2) being able to position logs of individual players in contrast to images. As a result, our proposed model is able to (1) provide the overall scenario in terms of the possible results of the business game, (2) show the relationship between the results of each individual player and the overall scenario and simultaneously presents other results of "possible" of the player, and (3) analyze the difference factors among players. Thus, it provides useful information for both facilitators and players during debriefing.

When evaluating player's behavior in a business game, it is common to compare and analyze individual play-logs by human players. In addition, there have been attempts to improve the feasibility of game execution and the decision-making model by replacing some human players with virtual agents. However, our proposed method takes the opposite approach to extending to business games in a way that human players can participate from ABM as a starting point. With numerous trials and simulation analysis method by ABM it is possible to judge the characteristics of the player from a small number of individual logs and it becomes possible to provide the evaluation axis to the less experienced facilitator and the newly created game.

Future studies should target business games in which large numbers of human players are required to make a series of decisions. In terms of feasibility, we restricted this study to cases in which one human player made a single decision, but it is possible to deal with multiple players and continuous decision-making. In addition, we adopted the log clustering method to type the scenario/simulation path. However, as arbitrariness also exists, such as in relation to the number of clusters, other methods should also be considered.

References

1. Arai, K., Deguchi, H., Kaneda, T., Kato, F., Nakamura, M.: Gaming Simulation. Union of Japanese Scientist and Engineers (1998). (in Japanese)
2. Koshiyama, O., Kunigami, M., Yoshikawa, A., Terano, T.: Analyzing behaviors of business game learners using a modified performance sheet. Stud. Simul. Gaming **21**(2), 86–95 (2011). (in Japanese)
3. Terano, T.: Learning business decisions through cases and games. J. Soc. Instrum. Control Eng. **46**(1), 44–50 (2007). (in Japanese)
4. Fujimori, H.: Alexander Islands. University of Tsukuba Graduate School of Business Science, Otsuka, Bunkyo-ku, Tokyo (1999)
5. U-Mart WEB page. http://www.u-mart.org/html/. Accessed 20 Feb 2018
6. Inoue, A.: Gamification. NHK Publishing (2012)
7. Terano, T., Koyama, Y.: Gamification: designing the world as games. J. Soc. Instrum. Control Eng. **54**(7), 494–500 (2015)

8. Nakano, K., Matsuyama, S., Terano, T.: Research on a learning system toward integration of case method and business gaming. In: Proceedings on the 4th International Workshop on Agent-based Approach in Economic and Social Complex Systems (AESCS 2007), pp. 21–32 (2007)

9. Kobayashi, T., Takahashi, S., Kunigami, M., Yoshikawa, A., Terano T.: Is there innovation or deviation? Analyzing emergent organizational behaviors through an agent based model and a case design. In: Proceedings on the 5th International Conference on Information, Process, and Knowledge Management (eKNOW 2013), pp. 166–171 (2013)

10. Goto, Y., Sugimoto, A., Takizawa, Y., Takahashi, S.: Methodology for facilitating understandings of complex agent-based models by gaming. Trans. Inst. Syst. Control Inf. Eng. 27(7), 290–298 (2014). (in Japanese)

11. Tanaka, Y., Kunigami, M., Terano, T.: What can be learned from the systematic analysis of the log cluster of agent simulation. Stud. Simul. Gaming 27(1), 31–41 (2017). (in Japanese)

12. Shin, H.S.: Risk and Liquidity, 1st edn. Oxford University Press, Oxford (2010)

13. Kikuchi, T., Kunigami, M., Yamada, T., Takahashi, H., Terano, T.: Analysis of the influences of central bank financing on operative collapses of financial institutions using agent-based simulation. In: IEEE the 40th Annual International Computers, Software & Applications Conference, The 3rd International Workshop on Social Services Through Human and Artificial Agent Models (2016)

14. Kuroda, H., Kato, I.: Tokyo Money Market. Totan Research ed., 7th edn., Yuhikaku (2009) (in Japanese)

15. FSA WEB page. http://www.fsa.go.jp/policy/basel_ii/basel3.pdf. Accessed 20 Feb 2018

16. Luenberger, D.G.: Investment Science. Oxford University Press, New York (1997)

Applications of Evolutionary Computation and Artificial Intelligence in Metallurgical Industry

Jianqi An[1,2,3]([✉])(iD), Jinhua She[1,2,3](iD), Huicong Chen[1,2](iD), and Min Wu[1,2](iD)

[1] School of Automation, China University of Geosciences, Wuhan 430074, China
{anjianqi,wumin}@cug.edu.cn, 1070037900@qq.com
[2] Hubei Key Laboratory of Advanced Control and Intelligent Automation
for Complex Systems, Wuhan 430074, China
[3] School of Engineering, Tokyo University of Technology, Tokyo 192-0982, Japan
she@stf.teu.ac.jp

Abstract. Metallurgical industry is one of the most important industrial processes, which mainly consists of coking process, sintering process, ironmaking process, and casting and rolling process. All of the metallurgical processes are complex, multivariate and nonlinear systems with large time-delay. Some chemical or physical mechanisms are even not clear and uncertain. It is difficult to establish the models, design the controllers, devise the scheduling and optimization strategies, and make the operation decisions by the conventional mechanism-based methods. Nevertheless, these processes work continuously and repetitively, which produces large amounts of data, and consists of lots of knowledge and expert experiences. In the last decade, evolutionary computation and artificial intelligence (ECAI) began to be widely used in metallurgical industry and many good results were reported. This letter demonstrates how the development of ECAI impacts the metallurgical industry by analyzing some good applications of the ECAI in typical metallurgical processes and discusses the future development trends and challenges of the applications of the ECAI in metallurgical industries.

Keywords: Evolutionary computation · Artificial intelligence · Metallurgical industry

1 Introduction

Evolutionary computation and artificial intelligence (ECAI) has developed rapidly in recent decades and has become a cross-disciplinary subject with a wide range of topics. With its big advantages in processing for large-scale data and optimization for complex problems, ECAI has been effectively used in various industries and achieved a lot of good application results. Metallurgical industry is one of these industries. Almost all the metallurgical processes are complex, multivariate and nonlinear systems with large time-delay. Some chemical or physical

© Springer Nature Singapore Pte Ltd. 2019
F. Koch et al. (Eds.): GEAR 2018, CCIS 999, pp. 77–87, 2019.
https://doi.org/10.1007/978-981-13-6936-0_9

mechanisms are even not clear and uncertain. It is difficult to establish the models, design the controllers, devise the scheduling and optimization strategies, and make the operation decisions based on conventional mechanism-based methods. In the last decade, ECAI began to be widely used in metallurgical industry and many good results were reported [1].

Metallurgical industry, in particular, the iron and steel industry, is one of the most important industrial processes. It converts various iron-minerals (oxides, sulfides, hydrated ores, silicates, carbonates, etc.) to metals by a serious of complex chemical reactions and physical processes under high-pressure and high-temperature conditions, which consumes great energy and produces lots of air and water pollution. Conventional modeling, control, scheduling, and optimization methods in metallurgical processes are mainly based on the analysis of the reaction mechanisms, the theory of material and heat balance, thermodynamics, dynamics, and other production mechanisms.

For example, in a coal blending process, which is to blend different kinds of coal to make coke, the ratio of the different kinds of coal affects the quality and price of coke [2]. Conventional manual calculation methods usually cost a long time and cannot accurately predict the quality of the coke, thus it is difficult to obtain an optimal blending ratio of coals to yield qualified coke with the lowest price. And in a blast furnace (BF), the temperature field of the burden surface is an important factor in judging the gas-flow distribution in the throat of the BF and the running condition of the BF [3]. However, there are no directly detecting devices for the temperature field. Conventional methods are to estimated by operators based on expert experiences, which has low accuracy and efficiency.

Due to the complexity and uncertainty of the metallurgical process as well as the limitation of the understanding and analysis of processes, it is very hard to design accurate process models, controllers, and optimizers based on conventional mechanism-based methods. In the recent years, under the background of high demands on energy conservation and emission reduction, fuzzy control [4], artificial neural networks (ANN) [5], expert systems, pattern recognition, genetic algorithms (GA) [6], particle swarm optimizations (PSO) [7], and many other ECAI technologies have been developed and applied to metallurgical processes to improve the quality and efficiency, reduce the energy consumption and pollution emission, and improve benefits of enterprises.

The metallurgical industry mainly consists of coking process, sintering process, ironmaking process, steelmaking process, combustion process, casting process, and rolling process. All of them are complex, multivariate and nonlinear systems with large time-delay. The rest of this letter reviews some successful applications of ECAI in some typical metallurgical processes, analyzes some typical ECAI systems, and discusses the future development trends and challenges of the applications.

2   Applications of ECAI in Sintering Process

Sintering is a process of compacting and forming a solid iron ore by heat without melting it to the point of liquefaction. It is an effective way to reduce porosity and enhance the properties of iron ore, such as strength, electrical conductivity, translucency, and thermal conductivity. In the process, iron ore, solvents, fuels, and sintering recycled materials are laid on a sintering trolley and sintered in a certain proportion. The conventional mechanism methods have solved the problems of total concentrate control, granulation humidity control, sinter-layer thermal-state modeling, and other problems. However, sintering is a very complex process that contains uncertainty, strong coupling, nonlinearity, and hysteresis. And many key steps in the sintering are still controlled based on expert experiences which depend on the actual experience of operators and personal prediction ability. Therefore, it is difficult to adapt to the changes in working conditions. Thus, the quality of the sinter is fluctuating.

Intelligent methods have been developed during the past few years for sintering processes. Some of them have been successfully applied to the actual metallurgical process and yielded good results. In the proportioning of iron ore, different kinds of iron ores with coke, limestone, dolomite, and returned sinter are mixed to produce a raw mix for the production of qualified sinter. The chemical components, proportions of the raw materials, and the price of the raw materials need to be considered as optimizing objectives. Regarding the problem, an intelligent integrated optimization system was developed for the proportioning, which contains optimization of the first proportioning, optimization of the second proportioning, a cascade integrated quality-prediction model, and a GA-PSO algorithm [8]. A multi-objective optimization and an analysis model of the sintering process based on GA-ANN were designed to optimize the production cost, energy consumption, and quality simultaneously [9]. In addition, burning-through point (BTP) control is another important issue which is the key to guaranteeing the sintering quality. Thus, many new ECAI methods have been designed for modeling and control the BTP. For example, an improved GM(1,1) model of grey system theory was designed to predict the sintering state parameters [10]. In order to improve the accuracy of the soft measurement for BTP, a cubic curve fitting and a quintic curve fitting were adopted to detect the BTP on both sides of a sintering machine [11]. An intelligent integrated optimization and control system were presented for the lead-zinc sintering process [12]. A two-fuzzy-controller coordinating control method was designed to control the BTP and bunker level [13]. An improved Takagi-Sugeno (T-S) fuzzy model of BTP based on a linear parameter varying and a fuzzy robust control method was proposed to control the BTP accurately [14].

The above-mentioned intelligent methods solved many problems that are difficult for conventional methods. We take the prediction and control of BTP as an example to explain the application of ECAI which is shown in Fig. 1 [15]. This system contains five parts: a soft measurement model which feeds back the actual BTP [16]; a prediction model which predicts the BTP by the sintering trolley speed, ignition temperature, bellow exhaust temperature, and mixed moisture [17]; a hybrid fuzzy-predictive controller which controls the BTP by adjusting the trolley speed and ignition temperature based on the fuzzy theory [18]; a level expert controller which produces control output based on expert system;

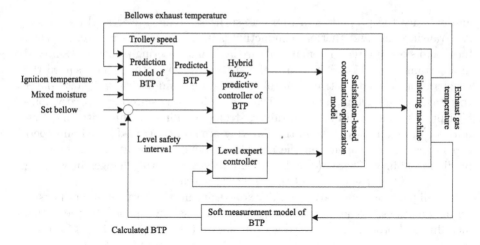

Fig. 1. Block diagram of intelligent control system for sintering end point process.

and a satisfaction-based coordination optimization model which is designed for yielding the optimal control by comparing the outputs of the fuzzy controller and expert controller for the sintering process [19]. Real-world applications show that this system combined with several ECAI technologies successfully controlled the BTP stable around a given bellow.

3   Applications of ECAI in Coking Process

In a coking oven, different kinds of coal are carbonized to coke by heating in the absence of air so as to expel the volatile ingredients. The coking process is a complex heat transfer and chemical change process. Iron and steel enterprises require comprehensive production targets with the largest coke production and the lowest energy consumption under the premise that the coke quality meets the requirements of iron-making in a BF. Due to the complicated structure of the coke oven, harsh operating environment, and few detection devices, conventional methods mainly rely on manual experience. Thus, it is difficult to adjust the coking process in real time to adapt to the complicated environmental vibrations.

Many intelligent methods were proposed to solve some of the problems well with their powerful computing optimization ability. Regarding the coal blending, the relationships between basic parameters of different coals and the microtexture of their produced cokes have been studied based on ECAI methods. For example, a multi-objective GA was designed to achieved optimal blending ratio of the different kinds of coals in order to get maximum yields and minimum input cost of raw coal [20]. And an integrated method combining with an ANN and a simulated annealing algorithm was presented to optimize the blend ratio [21]. Regarding the combustion process, a relationship model between the process parameters and the coke-oven temperature was designed based on an active semi-supervised affinity propagation clustering and a least-squares support vector machine (SVM) [22]. An integrated soft-sensing method was presented to estimate the coke-oven temperature by integrating linear regression and supervised distributed neural network models, an expert coordinator, and a model adaptive unit [23]. And a fuzzy controller with online-optimized parameters was designed to guarantee the control performance of the fuzzy control system for the combustion process in a coke oven [24]. Regarding other processes, a hierarchical intelligent decoupling control method for the gas collection process of three asymmetric coke ovens was successfully applied to sufficiently suppress disturbances and accurately stabilize the pressures [25]. And an integrated approach by combining an improved GA with a T-S fuzzy model was designed for modeling the oxygen content in a coke oven [26].

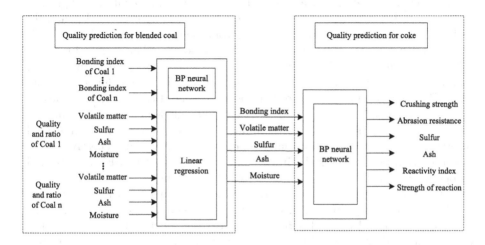

Fig. 2. A two-step intelligent prediction method for coke quality.

We take the optimization of the blending ratio of different types of coal as an example to explain an application of ECAI in the coking process. Since the optimization objectives of the coal blending process are complex and there are many variables and constraints, it is a computationally intensive task to find an optimal ratio of the different types of coal, which is difficult for conventional methods. Therefore, some effective ECAI methods were designed to solve this problem. The basic idea of the methods mainly contains three parts: a prediction model for blending-coal quality, a prediction model for coke quality, and a optimization algorithm for calculating blending ratio considering the predicted quality of the coke and the price of the input coals (The quality and price of different types of coal are not the same). These methods find out the relationship between the quality index of coke and the quality index of coals based on historical production data and optimize the blending ratio for maximum yields and minimum input cost of raw coals under constraints of the coke quality.

Figure 2 shows a two-step intelligent prediction method for coke quality [22]. It first combined a BP neural network and a linear regression to predict the quality of the blending coal based on the properties of each raw coal. Next, a BP neural network predicted the coke quality based on the quality of the blending coal. Based on the predicted coke quality, [22] also presented a simulated annealing algorithm to find an optimal blending ratio for maximum yields of coke and minimum input cost of raw coal under constraints of the coke quality. The results of a real-world application show that this method accurately predicted the coke quality and effectively reduced the cost of the coke.

4   Applications of ECAI in Ironmaking Process

A BF is a complex metallurgical reactor that converts iron ore into liquid pig iron through a series of chemical reactions and physical changes. In a BF, the coke and iron ore are burdened into the furnace from the top, and the preheated air is blown into the furnace through tuyeres from the bottom. The chemical reaction takes place throughout the furnace as the solid materials descend and the air goes up. It consumes a very large amount of energy and yields liquid hot iron, liquid slag, and a large amount of hot gas consisting of carbon dioxide, carbon monoxide, nitrogen, and hydrogen. It is very difficult for conventional method to model and control BFs due to the limited detected parameters, complicated reactions, and unclear mechanism. Nowadays, although some effective ECAI methods were proposed for the ironmaking process, most of BFs are adjusted by operators, which highly depends on expert experiences.

Intelligent methods avoid to deeply analyze the complex mechanisms of ironmaking to a certain extent, which solved many problems effectively. For example, a decoupling control method with fuzzy theory was devised to control the top pressure of a BF [27]. Regarding the estimation of state parameters, a two-stage online prediction method based on an improved echo state network [28], a quantile regression-based echo state network ensemble [29], and a least-square SVM model based on online hyper-parameters optimization [30] were proposed to forecast BF gas flow, respectively. A soft-sensing method was designed for detecting the slag-crust state of a BF based on two-dimensional decision fusion [31]. A Bayesian block structure sparse based T-S fuzzy modeling method [32] and a integrate method using support vector regression (SVR) combined with clustering algorithms [33] were proposed to predict the hot metal silicon content, respectively. And a principal component analysis algorithm was developed to monitor the iron-making process and achieved early abnormality detection [34]. Regarding the decision for operations, a fuzzy-based SVM multiclassifier was proposed to provide a more direct indication for operators to control the hot metal silicon content of a BF [35]. A data-driven prediction model was presented to adjust the burden distribution matrix based on an improved multi-layer extreme learning machine algorithm [36]. And an intelligent decision-making strategy for determining the burden distribution parameters was devised for improving energy-consuming index [37].

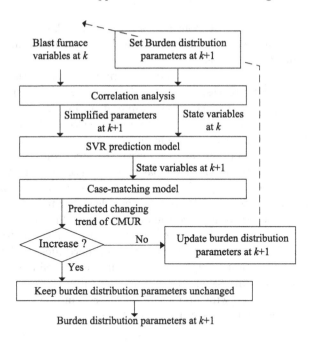

Fig. 3. Intelligent decision-making strategy for burden distribution of a BF.

We take an intelligent decision-making strategy for burden distribution of a BF (Fig. 3) as an example to explain the application of ECAI in the ironmaking process [37]. Considering the operators are more concerned about the changing trend of carbon-monoxide utilization rate (CMUR) rather than its instantaneous value, the CMUR is taken as an energy-consuming index. This strategy includes two models. One model is a prediction model based on an SVR which takes state variables at time k and burden distribution parameters at time $k + 1$ as inputs, and the predicted state variables at time $k+1$ as output. The other model uses the predicted state variables to determine the changing trend of CMUR by a probability-based case-matching model. The case-matching model gives not only the predicted changing trend of CMUR but also the probability of its occurrence. If the increasing probability of CMUR is higher than its decrease, then the burden distribution parameters are suitable and kept unchanged. Otherwise, the burden distribution parameters need to be updated. Simulation results based on industrial data show that the decision-making strategy provides a good guide on making a suitable decision for burden distribution parameters.

5   Applications of ECAI in Steel Rolling Process

The steel rolling process is a metal forming process in which metal stock is passed through one or more pairs of rolls to reduce the thickness and to make the thickness uniform. The rolling production consists of heating, rolling, annealing, and other processes. It is a fast, continuous process with large disturbance and complex dynamic nonlinear characteristics. It is hard to achieve satisfied control results by conventional methods.

Many ECAI methods were proposed and applied to solve these problems. Regarding the heating process, an integrated intelligent decoupling control method was designed to adjust the zone temperature in a regenerative pusher-type reheating furnace [7]. The intelligent decoupling control combines a fuzzy neural network with a hybrid PSO scheme. Regarding the rolling process, a hybrid PID controller was designed to improve the steel rolling process performance by dynamic adjusting the PID parameters based on an ANN method [38]. An ANN method was applied to improve the model prediction ability for rolling force and rolling torque [39]. A long-term learning method using a neural network was proposed to improve the accuracy of rolling force prediction in hot-rolling mill [40]. And a soft sensor was developed for the hot-steel rolling mill process by using least-squares SVM and a properly designed bias update term [41]. Regarding the annealing process, a dynamic modeling method for the quality control of a real large-scale continuous annealing process was proposed, which used a generalized growing and a pruning RBF neural network to establish the required dynamic quality control model [42].

6   Conclusion

This letter showed the development of ECAI methods in metallurgical industries by reviewing and analyzing some successful application examples in the past decades. Many of these ECAI methods achieved good results in real-world applications with their advantages in processing large-scale data and optimization for complex problems. They either improved production efficiency, reduced cost, cut down the labor intensity of operators, increased the level of the automation and information, or reduced the pollution emissions.

With the increasing requirements for environmental protection and energy resources, the metallurgical industry is facing greater difficulties and higher requirements. In the future, ECAI methods will be used more and more widely in metallurgical industries. The performance requirements for the ECAI methods will be much higher and various different types of ECAI systems will be developed. For example, the integration of rules, models and framework technologies with intelligent methods will be used to establish integrated systems; big data technology and deep-learning approaches for the continuous metallurgical industry will draw a lot of interest; network-based and cloud-computing-based intelligent metallurgy systems and intelligent metallurgical robots will gradually be applied to the metallurgical industry.

However, there are still many challenges in the development of ECAI in metallurgical industries. Although the intelligent method can effectively avoid to deeply analyze the complex mechanisms of metallurgical processes, it is still impossible to achieve good results in the metallurgical industry by simply using ECAI methods. How to design a specific ECAI method based on comprehensively understanding of the complex mechanism of a metallurgical process is the biggest challenge for AI-engineers. In addition, although many applications of ECAI methods have been successfully applied in metallurgical processes, there are still many difficult problems cannot be solved by the presented ECAI methods. New ECAI methods need to be developed for metallurgical industries, such as the methods with large-scale, fast-running, high-efficiency, and deep learning capability. There are still a lot of problems in metallurgical industries that need to be addressed by developing effective ECAI methods.

Acknowledgement. This work is supported by Hubei Provincial Natural Science Foundation of China under Grants 2016CFB480 and 2015CFA010, National Natural Science Foundation of China under Grants 61333002 and 61203017, the Foundation Research Founds for China University of Geosciences under Grant 2015349120, and the 111 project under Grant B17040. The first author is an overseas researcher under Postdoctoral Fellowship of Japan Society for the Promotion of Science (JSPS), and his JSPS Fellowship ID is P16799.

References

1. Chertov, A.D.: Application of artificial intelligence systems in metallurgy. Metallurgy **7**, 32–37 (2003)
2. Yin, C., Luo, Z., Zhou, J., et al.: A novel non-linear programming-based coal blending technology for power plants. Chem. Eng. Res. Des. **78**(1), 118–124 (2000)
3. Xie, N., Cheng, S.: Analysis of effect of gas temperature on cooling stave of blast furnace. J. Iron Steel Res. **17**(1), 1–6 (2010)
4. Martín, R.D., Obeso, F., Mochón, J., et al.: Hot metal temperature prediction in blast furnace using advanced model based on fuzzy logic tools. Ironmaking Steelmaking **34**(3), 241–247 (2007)
5. Bilim, C., Ati, C.D., Tanyildizi, H., et al.: Predicting the compressive strength of ground granulated blast furnace slag concrete using artificial neural network. Adv. Eng. Softw. **40**(5), 334–340 (2009)
6. Cierpisz, S., Heyduk, A.: A simulation study of coal blending control using a fuzzy logic ash monitor. Control Eng. Pract. **10**(4), 449–456 (2002)
7. Liao, Y., She, J., Wu, M.: Integrated hybrid-PSO and fuzzy-NN decoupling control for temperature of reheating furnace. IEEE Trans. Industr. Electron. **56**(7), 2704–2714 (2009)
8. Li, M., Wang, Q., Sun, Y.: Sintering blending optimization based on hybrid particle swarm algorithm. Inf. Control **37**(2), 242–246 (2008)
9. Zhang, J., Xie, A., Shen, F.: Multi-objective optimization and analysis model of sintering process based on BP neural network. Int. J. Iron Steel Res. **14**(2), 1–5 (2007)
10. Wu, M., Chen, X., Cao, W., et al.: An intelligent integrated optimization system for the proportioning of iron ore in a sintering process. J. Process Control **24**(1), 182–202 (2014)
11. Kim, B.R., Jeong, J.W., Hwang, K., et al.: Estimation of burn-through point in the sinter process. In: Proceeding of 14th International Conference on Control. Automation and Systems, pp. 1531–1533. IEEE, South Korea (2014)
12. Wu, M., Xu, C., She, J., et al.: Intelligent integrated optimization and control system for lead-zinc sintering process. Control Eng. Pract. **17**(2), 280–290 (2009)

13. Xiang, J., Wu, M. Duan, P., et al.: Coordinating fuzzy control of the sintering process. In: Proceeding of 17th IFAC World Congress, pp. 7717–7722. Elsevier, Seoul (2008)

14. Chen, X., Hu, J., Wu, M., et al.: T-S fuzzy logic based modeling and robust control for burning-through point in sintering process. IEEE Trans. Industr. Electron. **99**, 9378–9388 (2017)

15. Wu, M., Cao, W., Chen, X., et al.: Intelligent optimization and control of complex metallurgical processes. Springer, in pressing

16. Chen, X., Chen, X., She, J., et al.: A hybrid just-in-time soft sensor for carbon efficiency of iron ore sintering process based on feature extraction of cross-sectional frames at discharge end. J. Process Control **54**, 14–24 (2017)

17. Wang, C., Wu, M.: Hierarchical intelligent control system and its application to the sintering process. IEEE Trans. Industr. Inf. **9**(1), 190–197 (2012)

18. Wu, M., Duan, P., Cao, W., et al.: An intelligent control system based on prediction of the burn-through point for the sintering process of an iron and steel plant. Expert Syst. Appl. **39**(5), 5971–5981 (2012)

19. Xiang, L., Wu, M., Xiang, J.: A fuzzy sliding Model Control Strategy for the Burning through point and its application in sintering process. J. East China Univ. Sci. Technol. **32**(7), 820–836 (2006)

20. Chakraborty, A., Chakraborty, M.: Multi criteria genetic algorithm for optimal blending of coal. Opsearch **49**(4), 386–399 (2012)

21. Deng, J., Lai, X., Wu, M., et al.: Intelligent optimization method for coal blending based on neural network and simulated annealing algorithm. Metall. Ind. Autom. **31**(3), 19–23 (2007)

22. Lei, Q., Yu, H., Wu, M., et al.: Modeling of complex industrial process based on active semi-supervised clustering. Eng. Appl. Artif. Intell. **56**, 131–141 (2016)

23. Wu, M., Lei, Q., Cao, W., et al.: Integrated soft sensing of coke-oven temperature. Control Eng. Pract. **19**(10), 1116–1125 (2011)

24. Lei, Q., Wu, M., She, J.: Online optimization of fuzzy controller for coke-oven combustion process based on dynamic just-in-time learning. IEEE Trans. Autom. Sci. Eng. **12**(4), 1535–1540 (2015)

25. Wu, M., Yan, J., She, J., et al.: Intelligent decoupling control of gas collection process of multiple asymmetric coke ovens. IEEE Trans. Industr. Electron. **56**(7), 2782–2792 (2009)

26. Zhang, R., Tao, J., Gao, F.: A new approach of takagi-sugeno fuzzy modeling using an improved genetic algorithm optimization for oxygen content in a coke furnace. Industr. Eng. Chem. Res. **55**, 6465–6474 (2016)

27. An, J., Yang, J., Wu, M.: Decoupling control method with fuzzy theory for top pressure of blast furnace. IEEE Trans. Control Syst. Technol. https://doi.org/10.1109/TCST.2018.2862859

28. Zhao, J., Wang, W., Liu, Y., et al.: A two-stage online prediction method for a blast furnace gas system and its application. IEEE Trans. Control Syst. Technol. **19**(3), 507–520 (2011)

29. Lv, Z., Zhao, J., Liu, Y., et al.: Use of a quantile regression based echo state network ensemble for construction of prediction Intervals of gas flow in a blast furnace. Control Eng. Pract. **46**, 94–104 (2016)

30. Zhao, J., Liu, Q., Pedrycz, W., et al.: Effective noise estimation-based online prediction for byproduct gas system in steel industry. IEEE Trans. Industr. Inf. **8**(4), 953–963 (2012)

31. An, J., Zhang, J., Wu, M., et al.: Soft-sensing method for slag-crust state of blast furnace based on two-dimensional decision fusion. Neurocomputing **315**, 405–411 (2018)
32. Li, J., Hua, C., Yang, Y., et al.: Bayesian block structure sparse based T-S fuzzy modelling for dynamic prediction of hot metal silicon content in the blast furnace. IEEE Trans. Industr. Electron. **65**(6), 4933–4942 (2018)
33. Hua, C., Wu, J., Li, J., et al.: Silicon content prediction and industrial analysis on blast furnace using support vector regression combined with clustering algorithms. Neural Comput. Appl. **28**(12), 4111–4121 (2017)
34. Zhou, B., Ye, H., Zhang, H.F., Li, M.L.: Process monitoring of iron-making process in a blast furnace with PCA-based methods. Control Eng. Pract. **47**, 1–14 (2016)
35. Gao, C., Ge, Q., Jian, L.: Rule extraction from fuzzy-based blast furnace SVM multiclassifier for decision-making. IEEE Trans. Fuzzy Syst. **22**(3), 586–596 (2014)
36. Su, X., Zhang, S., Yin, Y., et al.: Data-driven prediction model for adjusting burden distribution matrix of blast furnace based on improved multilayer extreme learning machine. Soft Comput. **22**(11), 3575–3589 (2018)
37. Wu, M., Zhang, K., An, J., et al.: An energy efficient decision-making strategy of burden distribution for blast furnace. Control Eng. Pract. **78**, 186–195 (2018)
38. Zhang, J., Xie, A., Shen, F.: A hybrid intelligent system for PID controller using in a steel rolling process. Expert Syst. Applicat. **40**(13), 5188–5196 (2013)
39. Bagheripoor, M., Bisadi, H.: Application of artificial neural networks for the prediction of roll force and roll torque in hot strip rolling process. Appl. Math. Model. **37**(7), 4593–4607 (2013)
40. Lee, D., Lee, Y.: Application of neural-network for improving accuracy of roll-force model in hot-rolling mill. Control Eng. Pract. **10**, 473–478 (2002)
41. Shardt, Y.A.W., Mehrkanoon, S., Zhang, K., et al.: Modelling the strip thickness in hot steel rolling mills using least-squares support vector machines. Can. J. Chem. Eng. **96**, 171–178 (2018)
42. Li, S., Chen, Q., Huang, G.B.: Dynamic temperature modeling of continuous annealing furnace using GGAP-RBF neural network. Neurocomputing **69**, 523–536 (2006)

Evolutionary Computation and Artificial Intelligence for Business Transactions

Apostolos Gotsias[1,2]([⊠])[iD]

[1] Department of Business Administration,
University of the Aegean, 82100 Chios, Greece
agotsias@aegean.gr
[2] Regulatory Authority for Energy, 11854 Athens, Greece

Abstract. Business transactions are at the core of economic analysis as well as a prime area for research in Evolutionary Computation and for implementing scenarios in AI settings. The paper's focus is on the issue of synchronizing activities and coordinating the agents of the firm in a supply-chain environment. After briefly discussing the general coordination model, the Hourglass model, we present a mathematical model for achieving coordination inside the firm and show how the agents' activities are coordinated in a department, as well as across departments. The coordination model specifies the synchronization conditions by considering message travel times and product/support operational requirements. The conditions for achieving coordination and the relationships between operational and support departments are an original contribution in the economics of the firm. In the final part of the paper, we indicate how the coordination model's results can be utilized in a dedicated AI environment for studying economic relations among firm/market participants.

Keywords: Coordination · Synchronization · Economic transactions

1 Introduction

Business operations are characterized by the transformation of materials to final goods which are demanded by consumers. Success in a business endeavor is associated with ensuring that costs are kept at minimum and that the desired product reaches the consumer in the least time. The combination of moving in space (minimum expenditure of resources) and in time (maximum speed) is a problem that lies in the core of economic exchanges and its dimensions need to be modelled and analyzed. Kunigami and Terano [1] suggest that we should utilize methodologies for *"what and how"* to measure when we examine administrative/business operations. They suggest a structure for the analysis of business environments, separating them into the Macro (Industry level), the Meso (Firm level) and the Micro level (Individual agents). They also note

This paper is the outcome of fruitful discussions I had with Professors T. Terano, A. Yoshikawa and A. Orita in USA (Kennesaw), Japan (Tokyo) and Greece (Chios). I thank them for inspiring me to explore a novel for an economist research area. I am also grateful and thankful to two anonymous referees for their constructive comments that considerably improve the paper.

© Springer Nature Singapore Pte Ltd. 2019
F. Koch et al. (Eds.): GEAR 2018, CCIS 999, pp. 88–105, 2019.
https://doi.org/10.1007/978-981-13-6936-0_10

that business (product related) and administrative (support) operations consist of business processes which can be analyzed with agent-based experiments.

Gotsias [2] proposes the Hourglass framework of the firm and presents the structure of the firm's product related processes. The basic Hourglass mathematical model (Gotsias [3]) involves a structured analysis of the firm's product operations by stressing the coordination and the synchronization of its activities. The coordination of operational activities entails an analysis of firm processes that take into consideration both spatial and temporal dimensions. The spatial and temporal analysis of firm operations is based on the works of Ceapa [4, 5] and it is utilized in [3] to mathematically model the movement of product and information for the first time in the economics literature. A. Ceapa proposes a model, based on physical measurements, as an alternative method to derive the mathematical results of Special Relativity in Physics. Gotsias adopts his mathematical model to analyze in [3] the firm's physical environment and to introduce relevant economic notions and relations. Information flows (signals) take the form of messages carrying information on product quantity and characteristics, as well as the times of the product's departure and expected arrival. These messages are used as triggers by the subsequent operational departments along a supply-chain to notify their agents to start their assigned activities.

In this paper we present a framework for analyzing the business firm (meso level) and its internal environment (micro level). The Hourglass model is introduced briefly in Sect. 2, and its structure is presented so that the synchronized movements of product and information can be studied through successive supply-chain stages. In Subsects. 3.1 and 3.2, the coordination model in [3] is summarily presented and its implications are examined. In Subsect. 3.3, the coordination model is extended by introducing the support process in the firm's operations and we provide the mathematical model associated with these.

In the fourth section, we relate the model's results to product transport and to the provision of related support activities in the firm's supply-chain. In this part we introduce the organizational description of the firm's supply-chain as well as the flow of information between departments. Hull [6] examines pragmatic information flows in a business environment and proposes to analyze supply-chain exchanges with the help of a Data Flow Diagram. We utilize this tool to emphasize synchronization and coordination issues as the product is being transferred towards the consumer.

In the fifth section we outline how the coordination model's results can be utilized in an AI dedicated environment. Krolikowski et al. [7] examine business relations among agents in a production cycle utilizing Multi-Agent Systems (MAS). They propose a simple framework of economic behavioral rules and decision making, under limited resources and time constraints and simulate multiagent behavior that produces organizational patterns for various scenarios. Their scenarios produce self-organization patterns of interacting agents over time and specify the agents' spatial arrangements and mutual relations. Based on this paper's model, we suggest much simplified scenarios for analyzing business processes that are based on spatial and temporal considerations.

2 The Hourglass Model

In concord with the Kunigami-Terano suggestion, Gotsias [2] highlights the internal structure of the firm and the organization of its internal parts. The firm is coordinated by its top executive, the Peak Coordinator, and this person's (organizational unit's) goal is to rapidly provide the product (i.e., transporting and processing related activities in minimum time) and at a low final price (minimum cost of producing, storing, transporting and considering the cost of wasted resources/time) to the consumer.

As shown in Fig. 1, the firm engages in exchanges with agents in its internal (lower cone) environment. The participating agent organizations in the internal environment are the product operations and support departmental sub-organizations as well as the shareholder/owner and the resource exit/entry sub-organizations. Product operations are associated with producing and forwarding the product to the consumer. Functional support (Accounting, Finance, Procurement, Marketing and Pricing) are related to planning, scheduling, adjusting and data keeping activities that support the product's movement along the supply-chain. The shareholder/owner suborganization is concerned with securing the appropriate rate of return for the firm's investment. Resources (human, financial and technological) are brought in and/or replaced (entry/exit) if their performance is comparatively less than their alternatives. These sub-organizations are coordinated by the Peak Coordinator who is located at the apex of the (internal) firm organization and assesses the product's path along with its progression.

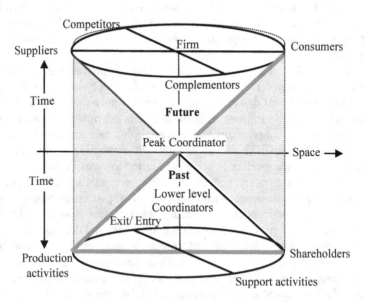

Fig. 1. The Hourglass model of the firm

The Peak Coordinator formulates and integrates the sub-organizations' plans and activities. The Peak Coordinator also represents the firm in the external environment (market, upper inverted cone) where it is in competition with other firms in the industry, relates with intermediate goods suppliers and cooperates with the firms that produce a complementary product to the firm (complementors) and is in rivalry with competitors.

3 The Product/Message Coordination Model

In this part we establish the existence of a unique coordinating point that is related to the moving speeds of a product and of a communicative message. A relation of their speeds determines the exact location of this coordinating point along a given distance, as shown in the next subsection. In the second subsection, we derive equations specifying the distances (times) that a firm must consider when it aims to satisfy consumer demand which is separated from its production facility by time and distance. In the third subsection we expand the model by adding the firm's support functions.

3.1 The Coordinating Point

Gotsias [3], based on Ceapa [4, 5], examines the coordination relationship that governs product and information (message) movements. We define a unit measure of distance (x) to study the synchronous movements of a product and its associated message. Given the speed of message transmission c, we have $t = x/c$ or $x = ct$, where t is the time the message takes to cover the unit distance x. In this time t, the product covers the distance vt (with v the product's speed of moving in space) when both the message and the product leave the origin of x, at $t = 0$. Thus, the remaining distance, till $x = ct$, is

$$x' = x - vt \text{ and } t' = t - \frac{vx}{c^2} \tag{1}$$

The left-hand side expression of Eq. (1) calculates the remaining distance x' as the difference of distances covered by the message and product movements. The right-hand expression calculates the time, t', in which a signal sent out from vt covers the distance x' and it is derived by dividing x' by c and utilizing $x = ct$. In the additional time $dt_1 = t_1 - t$, the signal at x is reflected (responded) back towards the origin and meets the forward moving product as shown by the top diagram of Fig. 2.

The reflected message meets with the forward moving product at point O_2 and we have

$$x' = vdt_1 + cdt_1$$

i.e., the remaining distance is covered by the reflected message and the product in time dt_1. Then t and dt_1 respectively are,

$$t = \frac{x'}{(c - v)} \text{ and } dt_1 = \frac{x'}{(c + v)}$$

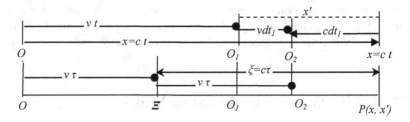

Fig. 2. The coordinating point Ξ, and the synchronized product-message movements

We define $\tau = (t + dt_1)/2$, as the average time of the total time required for the product to travel from O to O_2. It is shown in [3–5] that in the time τ, the product covers the distance $v\tau = O\Xi$ and that in the same time, the message covers the distance $c\tau = \xi = \Xi P$. Then (as can be seen on the lower diagram of Fig. 2), $x = v\tau + \xi$, or the unit distance x is completely covered by the moving product and message. In time 2τ, the product is at O_2 while the message is first reflected at P, and at time 2τ is back at Ξ. It is also shown in [3–5] that the values of ξ and τ are given by[1]

$$\xi = \beta^2 x', \quad \tau = \beta^2 \frac{x'}{c} \quad and \quad \beta = \frac{1}{\sqrt{1 - \frac{v^2}{c^2}}} \tag{2}$$

The distance ξ, covered by the message and $v\tau$, the distance covered by the product depend on the remaining distance to be covered by the product and on the coefficient β, which in turn depends on the relative product-information speed, c and v respectively (with c much larger than v). We refer to point Ξ as the coordinating point and Ξ is defined as the space point where product and message movements are synchronized, i.e., where they move in synchrony and that at every time interval τ, information and product coincide at a space point. The space-point Ξ of the synchronized movements depends on the product and message speeds and it is uniquely defined by them and for any distance separating the origin from the end-point. Distance ξ is the distance that the coordinating message must travel so that it reaches the end-point of x. In the following subsection we utilize the coordinating point when studying the product path from the production facility to the consumer. This path is both space and time dependent (the thick line in Fig. 1).

3.2 Coordination of Product and Message Movements

We perceive the consumer's position, designated by Q, at a distance $R = ct^*$ from its origin (ray of a circle with center) at O, as shown in Fig. 3. The ray is equal to distances $ct^* = OQ = OQ_1 = OP_1$, since all lie on the circle's periphery. We also note that the projection of Q on the horizontal axis is P_0 and that distance $OP_0 = OP'_0 = cT$. These distances are travelled by a message in time T, and they lie on the periphery of a circle

---

[1] Proofs are provided in the Appendix, parts 1, 2, 3 for τ, $x = v\tau + \xi$ and 2ξ, respectively.

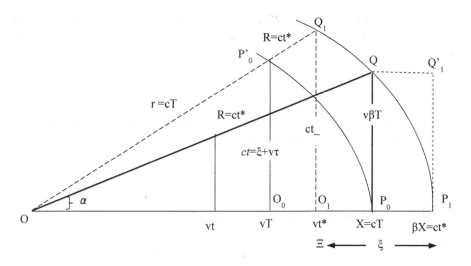

Fig. 3. Synchronized product and message movements from the Origin (O) to Demand point (Q)

also with center O, the origin. Our task is to calculate $R(t^*)$ and relate it to a coordinating point Ξ, which according to the previous discussion depends on time t [Eqs. (1), (2)]. From the triangle OO_1Q_1, utilizing the Pythagorean theorem and noting that $OQ_1 = ct^*$ and $O_1Q_1 = ct_-$, we have, [β defined in Eq. (2)]

$$t^* = \beta t_- \tag{3}$$

The time t^* required by a message to travel OQ_1, or by the product to travel OO_1, depends on the time t_- that is required by a message to travel the distance O_1Q_1. This message which is emitted from O_1 to Q_1 is perpendicular to the x-axis (space axis, where the product moves). We refer to this direction which is perpendicular to v's direction, as the *time-axis*. The time axis is used to record distances and to measure time intervals by a message which is sent out from the position reached by the moving product. When the product arrives at vT, a message that leaves the origin together with the product, arrives at $X = cT$ or, $OP_0' = OP_0 = cT = OQ \cos(\alpha)$. When the product is at vT, we lay the bottom diagram of Fig. 2 along the time axis O_0P_0' and we have on it, $x = ct = v\tau + \xi$. From the triangles OO_0P_0' and OO_1Q_1, we have

$$\frac{OP_0'}{OQ_1} = \frac{O_0P_0'}{O_1Q_1} \quad or \quad \frac{cT}{ct^*} = \frac{ct}{ct_-}$$

$$T = \frac{t^*\,t}{t_-} = \beta t \quad [by\ Eq.\ (3)] \tag{4}$$

Since $X = cT = OP'_0$ and $vT = OO_0$, by utilizing Eq. (4), we have $X = \beta ct = \beta x$ and $vT = v\beta t$. So, a new time dependent coordinate transformation is derived and related to Eq. (1),

$$X' = \beta(x - vt) \quad \text{and} \quad T' = \beta\left(t - \frac{vx}{c^2}\right) \tag{5}$$

Another form of the time equivalent transformation is also derived, since we have $x = \xi + v\tau$ and $t = \tau + v\xi/c^2$, when dividing the first expression by c. Thus,

$$X = \beta(\xi + v\tau) \quad \text{and} \quad T = \beta\left(\tau + \frac{v\xi}{c^2}\right) \tag{6}$$

Solving these equations for ξ and τ, we finally get

$$\xi = \beta X' = \beta(X - vT) \quad \text{and} \quad \tau = \beta T' = \beta\left(T - \frac{vX}{c^2}\right) \tag{7}$$

We look at the meaning of these equations now. Equation (5) is a simple transformation by which we have that the remaining distance x' is augmented by the β factor, along the x-axis, or $X' = \beta x'$. Thus, a message sent out from the origin at the same instant with the departing product, will reach $\beta x = X = P'_0$ and $v\beta t = vT$ respectively. The remaining distance X' has to be travelled by a message from O_0 to Q whose path, along the x-axis, is $O_0P_0 = O_0Q \cos(\theta)$ (not shown in Fig. 3). But the message travelling from O_0 to Q is not related to the message path from O to Q. These different paths imply that we have signal paths that are unrelated according to Eq. (5). Note that since T depends on t and t^*, then vT and X depend on t^*. But their difference depends on another signal as indicated before. Thus, we must find a way to relate O_0P_0 also on t^*.

Equation (6) is helpful in this task. It shows that X' equals the unit distance augmented by the β factor and it is shown along the time axis, perpendicular to vT. The left-hand side of Eq. (6) is equal to X when depicted on the x-axis. Equation (7) is another way of looking at Eq. (6), by solving it for ξ and τ. The left-hand side of Eq. (7) is denoted on the x-axis as the difference of $\beta X = ct^* = \beta^2 x$ and $v\beta T = vt^* = v\beta^2 t$ which equals $\xi = \beta X' = \beta^2 x'$. So, when the message and the product leave the origin at $t = 0$, they reach respectively $\beta X = c\beta T$ and $v\beta T$ in the same time, $t^* = \beta T = \beta^2 t$. The remaining distance, $\xi = \beta^2 x'$, is covered by a message in time $\tau = \beta^2 t'$, a time which concurs with the relations expressed in Eq. (1). Looking at the time-axis, we lay perpendicularly $\beta X = v\beta T + \xi$ on P_0 and we identify the vertical distance $P_0Q = v\beta T = v\beta^2 t$. Note that $ct = \xi + v\tau$ is perpendicular to vT and that the equivalent time transformation, Eq. (6), is depicted by $ct_- = c\beta t = X = \beta(\xi + v\tau)$ that is perpendicular to vt^*.

We have established a link between product and message movements that is governed by a relationship which depends on the speed of the product relative to the message's speed. This relationship, which is expressed by the factor β^2 times the difference in distances travelled by the message and the product, denotes a distance that the message should travel to meet with the forward moving product. This distance, defined by $\xi = \beta^2 x'$ identifies the coordinating point $\varXi$ and when divided by the message speed results in the time, τ, that is required for the message to cover this distance. The product is at the coordinating point $\varXi$ when the message is sent out to travel distance ξ in time τ and for the product to reach the intended position at Q or at P_1, it takes the additional time $\xi/v = \{(c/v) - 1\}t^*$. Thus, when the message reaches the demand point Q, the product is at $\varXi$, and by the time the product is at Q, in time ξ/v, the message travels the distance $c(\xi/v)$.

3.3   Coordination of Product, Support and Message Movements

In the previous subsection we studied the co-movement of the product and of a related message towards the consumer's location, where the good is demanded. But the product must be supported by services associated with its movement, i.e., procurement, transportation, storage services and administrative operations related to these (planning, communicating, record keeping, financing and accounting). The speed of transporting the product towards the consumer is affected by the speed of performing these associated operations along the supply-chain of the firm's product. In this subsection we examine the coordination of the moving product with the associated functional services.

We assume that the speed of providing support services along the firm's supply-chain is w and that $w > v$, i.e., support services can be provided at a faster rate per unit of time than the product's moving rate. We also assume that w is much smaller than the speed of the moving message, $c \gg w > v$. In the following sub-part, we relate the provision and processing of support activities to message movements. Our next task is to state the problem, analyze it and relate our results to the conclusions derived in the previous subsection.

Coordination of Support and Message Movements. It is obvious that a similar relationship should relate the speed of processing a support activity with the speed of the message, when both are initiated at $T = 0$. The analysis follows the steps of product movement in Sects. 3.1 and 3.2, with $ct = x$ and γ, the corresponding coefficient that relates processing and message speeds

$$\gamma = \frac{1}{\sqrt{1 - \frac{w^2}{c^2}}} \tag{8}$$

We also note that the results above correspondingly hold for times t, $\gamma T (=\gamma\beta t)$ and $\gamma^2 t$ just as they hold for t, βt $(=T)$ and $\beta^2 t$ $(=\beta T = t^*)$.

Co-movement of Product, Support Services and Messages. In this part we examine the situation when the product together with its associated documents and a message leave the origin at the same time. Our goal is to find a relation between these movements.

We depict the situation in Fig. 4, where we present two processes with different speeds of operation and three phases of analysis in three diagrams, an upper, a middle and a lower. Process A (product transport and storage) is related to speed v, process B (documentation and provision of support services) is associated with speed w and messages move with speed c. The upper diagram of Fig. 4 shows the space distances covered by processes A, B and by the message, at time T. The middle diagram shows how the top diagram of Fig. 4 is transformed if a message sent out from O_A reaches P_0 in time T', i.e., if the time required by the product to cover distance OO_A is not considered[2]. The middle diagram shows the remaining distance X' that the product must cover till P_0. This distance includes the distance that is due to the speed difference between product transport and support services and it is the same in the top and middle diagrams, $O_A O_B = O'_A O'_B$. If we consider the time (vT/c) it takes for the message to travel distance OO_A, we see that the message in this additional time (vT/c) arrives at P_2 and the support activities advance to O_{B2} in the top diagram, with $O_B O_{B2} = (wvT/c)$.

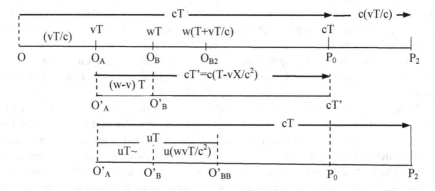

Fig. 4. Product and support speeds of movement and the addition of their relative speeds

The lower diagram in Fig. 4 presents the relative movements of A and B in time T. The message covers distance cT and in time T, some process of speed u, i.e., $uT \sim = (w - v)T$ covers the same distance $O'_A O'_B$ as denoted in the above two sub-diagrams. Then,

$$uT \sim = uT - u\left(\frac{wvT}{c^2}\right) \text{ and } T \sim = \left(1 - \frac{wv}{c^2}\right)T$$

or since $uT \sim = (w - v)T$ we have

$$u = \frac{(w - v)}{\left(1 - \frac{wv}{c^2}\right)} \tag{9}$$

Equation (9) calculates the relative speed of a process that takes place at O_B with respect to O_A. This relational process is initiated at O'_A in time T and reaches O'_B with

---

[2] Note that distance $O_A P_0 = X'$ according to Eq. (5)

speed u, in the associated time $T \sim$. As shown in the first part of the third subdiagram, the speed of support services relative to transporting operations, times the associated time $T \sim$, equals the difference of distances covered by the two processes in time T, $(w - v)T$. This is a governing, synchronized and coordinating relationship between any two processes and it is independent of the values of w and v.

Equivalent Time-Dependent Transformations. From the top diagram in Fig. 4, we have

$$X' = \beta(x - vt) \text{ and } T' = \beta\left(t - \frac{vx}{c^2}\right) \tag{10}$$

and if we assume $X = wT$, then (10) transforms in

$$X' \sim = (w - v)T \text{ and } T' = \mathrm{T}\sim = \left(1 - \frac{vw}{c^2}\right)T \tag{11}$$

We also have from the top diagram in Fig. 4,

$$X'' = \beta(x - wt) \text{ and } T'' = \beta\left(t - \frac{wx}{c^2}\right) \tag{12}$$

with X'' and T'', the remaining distance and the time in which a message covers the distance from O_B to P_0. Then, we also have,

$$X'' = (X' - uT') \text{ and } T'' = \left(T' - \frac{uX'}{c^2}\right) \tag{13}$$

and by using (9) in (13), considering (10) and (11), we derive Eq. (12). Thus, the above equations are equivalent, and they form a group.

4   Product, Support Activities and Message Coordination

In this section we study the synchronized product, support and message flows as the product is forwarded to the consumer. Since messages are characterized by much higher speeds than what products and/or support operations can attain, they are expected to arrive sequentially along the supply-chain's locations that grow farther from the origin and that are associated with different operations. Hence, the message's arrival in one location can be used as a signal to commence necessary local activities. Also, the message that arrives at one location can be used to notify the next one for any changes in the plan. For example, a message that travels to a given facility, faster than the product, allows enough time to notify the next facility to start preparing for the product's arrival. We study these issues next, by providing a general description of the process in the first subsection and analyzing it in terms of time requirements in the next subsection.

4.1   Product, Support and Message Movements

In Fig. 5, we employ a realistic Data Flow Diagram (DFD) [6] to display information flows in a supply-chain and their relation to synchronizing messages, product and support activities. Support activities refer to managerial decision making and planning that is related to scheduling services, account keeping and distributing information to interested parties. Scheduling matches items to be forwarded and transportation availabilities. Scheduling activities involve product related operational plans for loading/stocking/packing items on arrival and/or when outbound. When items are outbound, transportation must be available. Scheduling determines when items can be available for loading and triggers the dispatching of vehicles and of materials. Dispatching conveys formal orders to respective operations departments for releasing items and to the carrier for pickup. Dispatching ensures that vehicles and materials are available at the desired time and location.

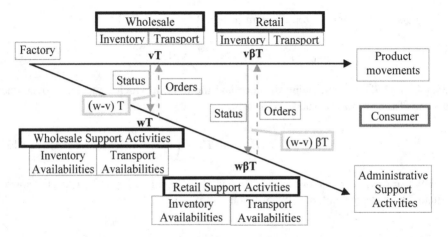

Fig. 5. Data flow diagram for a business supply-chain

We assume in Fig. 5 that a finished product and documents associated to production costs leave the Factory. At the same instant two messages leave, one towards Wholesale Operations and the other towards Wholesale Support. The messages advance to Wholesale, then to Retail and finally to the Consumer's respective locations. Each intervening supply-chain departmental unit is organized in two offices, one responsible for product related operations and the other for supporting the associated administrative activities. The Operations office is responsible for inventory and transport activities related to receiving, storing, loading/unloading and shipping the product. The Support office is responsible for ensuring that transportation and product inventories are available for forwarding it to the next location.

Two messages are sent out from each location, one internal to its associated office and the other external, to the next department down the supply-chain towards the consumer. The internal communication message informs the corresponding support office about the status of operations and the support office responds with the issuance of dispatching orders or updates the plan. The message for internal use is generated by the office the instant that the message originating from the Factory reaches the office. This external to the office message continues its path towards the consumer signaling to other intervening offices that the product has left the factory and its expected time of arrival. Close synchronization is required between product operations and support activities because scheduled activities need to be optimized to minimize wasted time.

4.2   Synchronizing Product, Support and Communication Flows

Figure 6 depicts an extended DFD of a business supply-chain and is supplanted with time duration data which relate to the advance of messages, product operations and support activities. We present the case of a push supply-chain (forward the product to the consumer) but the treatment of a pull supply-chain (the consumer requests the product) is analogous, acknowledging that the origin is now the consumer's location. We assume that the successive stages of the push supply-chain include the Factory (F), Warehouse (Wa), Wholesale (Wh), Retail (Rt), Delivery/Distribution (Di) departments and the Consumer location. Each department is organized into two offices, one responsible for Product Operations (Op) and the other associated with the department's Support (Sup) functions. Each department is at some distance from the Factory, along a path towards the consumer's location.

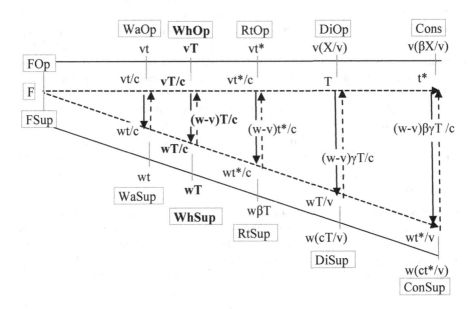

Fig. 6. Synchronization and coordination in a supply chain

In Fig. 6 the consumer is positioned at a distance ct^* away from the factory, also shown in Fig. 3 by OQ or by its distance equivalent OP_1. The projection of OQ on the x-axis is cT. In terms of Fig. 1, Production Activities refer to the Factory location and the Consumer is ct^* distance units away from the factory (heavy line). The heavy line connecting Production Activities and the projection on the x-axis (Shareholders) represents the product path along the x-direction (point P_0 in Fig. 3). Figure 6 is a bird's eye-view of the linkages between product, support and message paths. The analysis that follows is based on the interactions taking place at time T (shown in bold), and its extensions for times t, t^* are straightforward, since $t^* = \beta T = \beta^2 t$ with $T = \beta t$.

At time vT/c, a message arrives at $WhOp$ having left the FOp at $t = 0$, carrying information on the product volume and characteristics, as well as on T, the expected time of product arrival. At time vT/c, a status report from $WhOp$ is emitted towards $WhSup$ specifying the type and quantity of the arriving product, expected arrival, existing inventory and outbound capabilities. The notional distance between the offices is $(w - v)T$ and it takes $(w - v)T/c$ time for the message to cover this distance. When the intra-office message arrives at $WhSup$, the elapsed time since the message left the FOp totals $(vT/c) + (w - v)T/c = wT/c$, or it equals the time it takes for the $FSup$ office to communicate to $WhSup$ relevant information on product characteristics as well as financial and cost accounting information. At the same instant, two messages arrive at $WhSup$, one originating from $WhOp$ and the other from $FSup$ and after reconciling them, $WhSup$ issues a message with orders to dispatch the arriving product towards the next stage. The dispatch message arrives at time $(wT/c) - (w - v)T/c = vT/c$, or just at the time it takes for the message from $FSup$ to reach $WhSup$ plus the time it takes for the reflected message (negative sign, because of the change in direction) from $WhSup$ to $WhOp$. This cumulative time equals the time that is required for the product to arrive at $WhOp$ from FOp. Thus, as the product arrives, scheduling vehicles and dispatching orders arrive just in time and the product is shipped to the next supply stage. Inter-departmental messaging is used to synchronize successive departments of the firm, it allows for the coordination of their constituent offices and results in no wasted time in the firm's operations.

In conclusion, both the product and support related documentation arrive at Wholesale from the Factory, in time T, and in this time the internal exchange of messages is finalized, and intra-office coordination is attained. As was noted before, two messages are emitted from each office in the supply-chain. The intra-office messaging was examined above, and now we examine the structure of the inter-office message flows.

Inter-departmental Message Coordination. Since $cT = X$, support and transport activities are connected by the same distance X that the message covers in time T. Let us consider time T and record message/product/support movements. The remaining space distance to be covered by the transport activity is $X' = X - vT$, while for the support activity it is $X'' = X - wT$. Since X is the same for both equations, we have

$$X' = X'' + (w - v)T \tag{14}$$

The remaining space distance to be covered by the transport activity, X', is the sum of the remaining support space distance plus the difference between the activities' speeds times T. In terms of Fig. 6, X' is the distance separating $DiOp$ from $WhOp$, where the product has just arrived. This distance equals the sum of the intra-office distance in the Wholesale department $(w - v)T$ and of X'', the distance between $RtSup$ and $WhSup$[3].

But $X' = \xi/\beta$ and since (14) is true for every distance/time, like $\beta X'$ or $\gamma X'$, we have

$$\frac{\gamma}{\beta}\xi = \gamma X' = \gamma X'' + (w - v)\gamma T \tag{15}$$

and,

$$\gamma\xi = \gamma\beta X' = \gamma\beta X'' + (w - v)\gamma\beta T \tag{16}$$

Equation (15) specifies that the intra-office distance $(w - v)\gamma T$ [the distance between $DiOp$ and $DiSup$, or distance $(WhOp)$ to $(WhSup)$ augmented by γ] is the difference of the X' and X'' distances augmented by γ. Equation (16) relates the intra-office notional distance in the Retail department which results in the notional distance between Cons. and ConSup when augmented by γ.

The significance of Eq. (14) and its equivalents, Eqs. (15) and (16), is that by using them, the Peak Coordinator of the firm (or the departmental Coordinators) can monitor the evolvement of their operations and relate local operations to neighboring ones. The Coordinator is always located 2τ away from the distance that the message covered, as was seen in Fig. 2 and the related discussion there. In terms of Fig. 6, the coordinating relationships for times T, βT are respectively $2(\gamma/\beta)\xi$ and $2\gamma\xi$, considering the impact of Support activities. These are concrete relationships that hold for any neighboring processes and hold for any values of c, v and w. Thus, these equations should characterize any relations that are modelled inside a firm when neighboring agents transact.

5 Simulating Multi-agent Transactions

Krolikowski et al. [7] analyze economic relationships with their MASER (Multi-Agent System for Economic Relationships) model. The model is characterized by a structured environment where agents interact with neighbors, move and occupy a location in space. The MASER agents structure their behavior based on profit maximization subject to a survival constraint. The model distinguishes agents by traits who produce goods in specified periods (days), move according to their own speed, communicate with others (advertise) and keep records of their transactions. The MAS game is structured as follows: There are four agents (Farmer, Miller, Baker and Seller) randomly spaced from each other along the supply-chain (wheat, flour, bread, food). They are provided with some initial endowments (food, inventories) for production needs and they

---

[3] This is a special case used for exposition in Fig. 6. The distance X'' depends on the speed of Support Activities and varies according to the value of w.

communicate/advertise (sell messages) their product to the next stage's agents. Production of their good requires funds that are increased by selling the product and are reduced by the purchase of intermediate goods and food. The agents form relations (based on risk) and the stronger the relation, the closer they approach each other.

This paper's model resembles the MASER model but with some notable differences in terms of its structure. The model agents are linked to Warehouse, Wholesale, Retail and Seller departments and their positions are separated by distances determined by the speed of the moving product. Processing (production) times are given for each stage and specific times (distances) are required to reach the next stage. The duration of each activity is directly related to message transmission and processes are bound by their own speeds.

In MASER, the researcher sets the number of periods (days) in which production and exchanges are finalized. Also, the number of agents in each stage is set by the researcher without constraints, while this model's number of agents in each stage is constrained to two. The supply-chain's contracted agents, the department employee(s), form one group while the other is a group of competing agents who can perform each stage's functions, at a lower cost than the firm's provision. The last ones are included in the Exit/Entry sub-organization described in Fig. 1. Thus, competitive pressures from outside the firm ensure that costs are kept at minimum for each department. This model's agents receive an effort and food endowment (wages) either depending on the level of their contributions or not, and if their performance falls below the competitive threshold they are replaced. In the MASER model, the agent endowments are determined by the revenues (profits) accrued by sales and if these are not sufficient they exit the game.

Although the present analysis is in harmony with the MASER model, it may improve it by specifying consistent production times among agents, rather than unrestrictedly setting the time duration or "period". In MASER, the experimenter is free to choose the number of days for producing, transacting and forwarding the product to the next stage. Table 1, based on the preceding analysis, presents the agents involved, the space distances that separate them, as well as the message times that separate them. These parameter values can be used instead of the parameter "days" in MASER and the models' results can be evaluated. Based on the coordination model specified here, other traits can also be specified, like the agent's neighborhood ratio, the size of the market and the degree of competition between departments or offices. In short, the MASER model requires the setting of the parameters that specify number of agents, production times, subsistence cost, risk, trading neighborhood radius and the size of the market environment. In contrast to MASER, these parameters are endogenously determined in the coordination model and the researcher can experiment with different specifications and evaluate their importance.

Table 1. Distance separating departments and time for a message to cover it

	Remaining distance	Remaining time	Time difference
Factory	0	0	
dt			t'
Warehouse	$x' = x - vt$	$\dfrac{x'}{c} = t'$	
dt			$T' - t' = (\beta - 1)t$
Wholesale	$X' = (X - vT)$	$\dfrac{X'}{c} = T'$	
dt			$(\beta - 1)T'$
Retail	$\beta X' = \beta(X - vT)$	$\dfrac{\beta X'}{c} = \dfrac{\xi}{c} = \tau$	
dt			$\left(\dfrac{c}{v} - \beta\right)T'$
Seller/Delivery	$\dfrac{c}{v}X' = \dfrac{c}{v}(X - vT)$	$\dfrac{c}{v}\dfrac{X'}{c} = \dfrac{\xi}{v\beta}$	
dt			$\left(1 - \dfrac{1}{\beta}\right)\dfrac{\xi}{v}$
Customer	$\dfrac{c}{v}\beta X' = \dfrac{c}{v}\beta(X - vT)$	$\dfrac{c}{v}\dfrac{\beta X'}{c} = \dfrac{\xi}{v}$	

6   Conclusions

In [7], as in the traditional analysis of transaction exchanges, economic relationships are assumed to last for a "period" of some given duration, for example for a day or longer. Although it depends on the type of each transaction, a "period" is taken as a given according to traditional treatment in economics and is considered constant across successive transactions. But once a transaction's duration is not defined objectively, consecutive "periods" cannot easily be linked together since each "period" is of varying duration, depending on the type of transaction performed. Therefore, information transmission and message exchanges cannot be incorporated in the traditional analysis of transactions, since messages are assumed to travel between the transacting parties in this undefined time "period".

In this paper we propose a methodology for specifying the "time-period" in a consistent manner and which is associated with the agents' position in the supply-chain. This methodology is the outcome of the coordination mathematical model when considering the relevant conditions for attaining coordination and synchronization in business exchanges. Another contribution of the paper is that we extend the analysis in [3] by introducing the firm's support functions. Modelling business support functions, by considering jointly their spatial and temporal characteristics, constitutes an original contribution in the theory of the firm and it is advantageous in that it realistically depicts its daily business operations.

The modified, as suggested above, Multi-Agent System for Economic Relationships can be implemented in a dedicated environment and with it we can study patterns of self-organization inside and/or outside the firm's departments. With this modified MASER environment, with the use of artificial intelligence modelling and multiagent system analysis, the Message, Product and Support Activities Coordination model can be implemented and tested for both simple and complex scenarios, so that better decision-making can be attained at the level of the firm.

Appendix

1. Derivation of τ:

$$\tau = \frac{t + dt_1}{2} = \frac{\frac{x'}{c-v} + \frac{x'}{c+v}}{2} = \frac{\frac{(c+v)x' + (c-v)x'}{(c-v)(c+v)}}{2}$$

$$= \frac{2cx'}{2(c^2 - v^2)} = \frac{x'}{c\left(1 - \frac{v^2}{c^2}\right)}$$

Then, $\tau = \beta^2 \left(\frac{x'}{c}\right)$ and $\beta = \frac{1}{\sqrt{1 - \frac{v^2}{c^2}}}$

2. Derivation of $x = v\tau + \xi$:

$$x = ct = x' + vt = \frac{2(x' + vt)}{2} = \frac{x' + x' + 2vt}{2}$$

$$= \frac{(x - vt) + (vdt_1 + cdt_1) + 2vt}{2}$$

$$= \frac{ct + vt + vdt_1 + cdt_1}{2} = \frac{(c+v)t + (c+v)dt_1}{2}$$

$$= \frac{(c+v)(t + dt_1)}{2} = (c+v)\tau$$

Hence, $x = ct + c\tau = v\tau = \xi + v\tau$

3. Derivation of 2ξ: $OP + PO_2 = (c\tau + v\tau) + cdt_1$

and $\tau = \frac{t + dt_1}{2}$ or $dt_1 = 2\tau - t$

Hence, $OP + PO_2 = (c\tau + v\tau) + cdt_1$

$$= v\tau + \xi + c(2\tau - t) = v\tau + \xi + 2c\tau - ct$$

$$= x + 2c\tau - ct = x - x + 2c\tau$$

$$= 2c\tau = 2\xi$$

References

1. Kunigami, M., Terano, T.: Experiments based management and administrative science: a manifesto. In: The 1st General Conference on Emerging Arts of Research on Management and Administration (GEAR), Tokyo (2012)
2. Gotsias, A.: Experiments based management science and organizations. In: The 1st General Conference on Emerging Arts of Research on Management and Administration (GEAR), Tokyo (2012)
3. Gotsias, A.: A treatise on exchange coordination. Working Paper, University of the Aegean, Department of Business Administration, Chios, Greece (2015)
4. Ceapa, A.C.V.: Full physical derivation and meaning of Lorentz transformation (1999). https://arxiv.org/abs/physics/9911067v1. Accessed 7 July 2018
5. Ceapa, A.: Toward an exciting rebuilding of modern physics. In: Haranas, I. (ed.), p. 80 (2006). http://vixra.org/pdf/1004.0049v1.pdf. Accessed 7 July 2018
6. Hull, B.: A structure for supply-chain information flows and its application to the Alaskan crude oil supply chain. Logist. Inf. Manag. 15(1), 8–23 (2002)
7. Krolikowski, R., Kopys, M., Jedruch, W.: Self-organization in multi-agent systems based on examples of modeling economic relationships between agents. Front. Robot. AI 3(41), 1–16 (2016)

Proposal on Mutual Cooperation Between Simulation Research and Field Research in Archaeology

Fumihiro Sakahira(⊠)

KOZO KEIKAKU Engineering Inc., Nakano, Tokyo, Japan
f-sakahira@hotmail.co.jp

Abstract. In this paper, we propose a methodology to collaborate the research using agent simulation and the research using conventional method (field research) in archaeology. The main stream researches using Agent-Based Simulation (ABS) are unidirectional cooperation with the results of field research as input and the results of simulation research as output. In our proposed method, by presenting the hypothesis verification method from ABS result, the simulation result can become the input of field research. As an application example of proposed method, we discuss the problem of whether native Jomon people or Chinese-Korean immigrants played the major role of agricultural culture in Yayoi period by ABS.

Keywords: Agent-Based Simulation · Archaeology · Collaboration

1 Introduction

In this paper, we propose a methodology to collaborate the research using agent based simulation (ABS) and the research using conventional method (field research) in archaeology. As for the application examples, we will discuss the results of using ABS for "Who played the major role of Yayoi agricultural culture in Northern Kyushu?"

With the rapid increase in ABS research in archeology, there is concern about the lack of arrangements the lack of feedback between archaeologists and modelists [1]. One of the reasons for this is the deficiency of comprehensive textbooks and handbooks for archaeologists about simulation technique. In reference to this, a guide to the creation of a simulation model for archaeologists has been created [1]. Also, the complexity of software has been pointed out as an obstacle to the adoption of ABS by archaeologists. Therefore, it is mentioned that a software system that does not require advanced programming knowledge is required [2].

However, we do not believe that archaeologists will be able to collaborate on simulations and field studies because they learn to build models and create programs. We believe that in order to collaborate between simulated and field research, we should not link researchers together, but rather link research results together.

As an archaeological study using ABS, the most popular study is to examine the factors of demographic dynamics of ancient Anasazi people from 800 to 1350 in Long House Valley in Arizona, USA [3, 4]. These studies examined factors influencing the

© Springer Nature Singapore Pte Ltd. 2019
F. Koch et al. (Eds.): GEAR 2018, CCIS 999, pp. 106–111, 2019.
https://doi.org/10.1007/978-981-13-6936-0_11

population dynamics in Long House Valley by using several parameters including paleoenvironment variables from social unit and empirical data. Such researches are said to account for about 30% of all archeology studies using ABS [2]. That is, such researches can be said to be the mainstream of current research.

In these studies, scenario simulation is performed using historical facts based on anthropological and archaeological materials as input data, and in the simulation results according to historical facts, knowledge as to which parameter among the decision-making factors was effective is obtained. In other words, these researches are unidirectional cooperation with results of field research as input and results of simulation research as output.

2 Proposed Method

2.1 Concept

In this paper, we propose a method for bidirectionally linking research using ABS and field research.

The proposed method is the same as performing scenario simulation using historical facts based on anthropological/archaeological data as input data. However, it observes what kind of social phenomena occurred in the simulation results according to historical facts, generates a hypothesis about the problem. Moreover, it differs from these researches in that it presents what kind of archaeological materials at what time can verify the hypothesis. As an advantage of ABS, by attaching various attributes such as anthropological morphology, DNA, Food production system, culture etc. to the agent and observing the composition ratio of each attribute of the agent as the result of the scenario simulation in chronological order, based on the pattern of combination of attribute diffusion, it is possible to present materials that can verify the hypothesis. In other words, the proposed method could generate a working hypothesis that leads to the discovery of new remains and analysis of remains by presenting what kind of archaeological materials at what time can verify the hypothesis as the verification method of the hypothesis. That is, simulation can be input for research.

2.2 Procedure

The procedure of the proposed method is as follows.

1. Adopt a theme that seemingly inconsistent between materials corresponding to agent attributes.
2. In solving the above theme, consider necessary constituent elements at the minimum.
3. Create an agent model in which the rules are made based on the above and the inputs and constraints based on archaeological materials are attached as agent attribute variables.
4. Make cases by combining inputs based on archaeological materials and other variable parameters, and conduct simulation at many runs for many cases.
5. Extract the simulation case that satisfies the constraint condition.

6. If there are multiple cases with different parameters in the case, find the difference in the time series of the attribute variables of the agent between the cases.
7. Because attribute variables conform to archaeological data, the age and contents of the archaeological material are obtained for verifying which of the hypotheses are correct as the difference of the attribute in the time series of the simulation result.

3 Application Example

When applying the procedure of the proposed method to "Who played the major role of Yayoi agricultural culture in Northern Kyushu? [5]", it is as follows.

1. The theme about the major player of Yayoi agricultural culture in northern Kyushu was adopted as the theme of solving the conflict between the archaeological hypothesis that the native Jomon people played the major role of Yayoi agricultural culture and the anthropological hypothesis that the immigrants played the major role [6–9].
2. The above theme was defined as the problem about population increase due to the diffusion of agricultural culture and genetic, and as a component element, the food production system and trait genetic allele (Jomon trait and Immigrants trait) were set as the attribute variables of the agent.
3. The rule of diffusion of agricultural culture was made based on an infectious disease model, and the rule of inheritance of trait genes was made based on Mendel's law (Immigrants trait dominance).
4. 441 cases were created with variable parameters such as the propagation speed (range and introduction rate) of agricultural culture, and run ten times per case.
5. Of 441 cases, 111 cases could satisfy the constraint condition that 300 years after the initial state where a small number of immigrants migrated with agricultural culture to the place where many large numbers of native Jomon people were living, immigrants became the majority.
6. The 111 cases were divided into cases where the diffusion speed of agricultural culture is more slow and more fast. In each case, there was a large difference in the "trait gene" history of agent's attribute variable "agricultural culture" holder 50 years after the start (see Fig. 1). In the former cases, immigrants were the majority of agricultural culture holder. On the other hand, in the latter cases, native Jomon people were the majority of agricultural culture holder. On the problem about the major player of Yayoi agricultural culture, we generated the former hypothesis that immigrants played major role and the latter hypothesis that native Jomon people played major role.
7. Archaeological data corresponding to agents holding both agricultural culture and Jomon trait genes at the time of 50 years from the start, i.e., discovering of human bone showing Jomon trait and accompanying archaeological remain showing agricultural culture is an archaeological material that can verify which hypothesis is correct (Fig. 2).

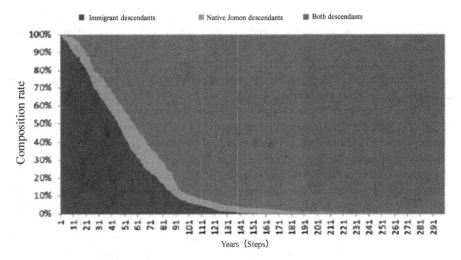

Fig. 1. Comparison on composition ratio of descendants of those practicing an agrarian culture in the cases of slow diffusion of the agrarian culture.

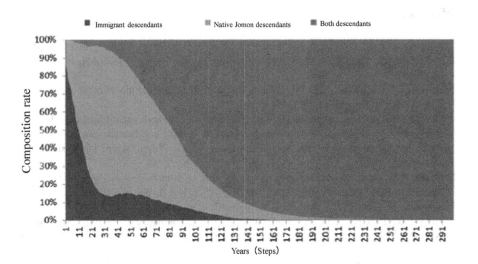

Fig. 2. Comparison on composition ratio of descendants of those practicing an agrarian culture in the cases of rapid diffusion of the agrarian culture.

4 Discussion

Firstly, we acknowledge important prior art in the field of ABS for human systems simulation such as Epstein and Axtell [10] seminal work on simulating social systems; Gilbert and Troitzsch [11] work on the use of simulation by social scientists; the work by Tesfatsion [12] on agent-based simulation for social economics, and; Bonabeau [13] on the utilization of agent-based modeling of human system.

In our work, based on the ABS results, we show work hypotheses that lead to the discovery of new remains and the reanalysis of remains for hypothesis verification, and by simulation being the input of field research, we show it made it possible to link simulation research and field research bi-directionally.

In the application example using the proposed method, it was possible to reproduce past social phenomena and generate a hypothesis by ABS. Then, for these hypotheses, by assigning various attributes (morphology, DNA, food production system, culture) to the agent in multiple, and seeing the component ratio of each attribute of the agent as a simulation result in chronological order, based on the pattern of combinations, we were able to present materials that can verify hypotheses.

Furthermore, we propose to analyze what kind of factors, that is, what rules and parameters are important for classification of many scenarios and many hypotheses. In addition, we think that it is necessary to analyze which microscopic phenomenon at any point in the simulation step is affecting the factors that result in different results even under the same condition. For these, it is effective to use the approximate Bayesian computation, the decision tree of simulation logs [14] and the random forest algorithm that can separate rules and parameter combinations.

References

1. Romanowska, I.: So You Think You Can Model? A Guide to Building and Evaluating Archaeological Simulation Models of Dispersals. Human Biology Open Access Pre-Prints 79 (2015)
2. Cegielski, W.H., Rogers, J.D.: Rethinking the role of agent-based modeling in archaeology. J. Anthropol. Archaeol. **41**, 283–298 (2016)
3. Dean, J.S., et al.: Understanding Anasazi culture change through agent-based modeling. In: Kohler, T.A., Gumerman, G.J. (eds.) Dynamics in Human and Primate Societies: Agent-Based Modeling of Social and Spatial Processes, pp. 179–206. Oxford University Press, New York (2000)
4. Axtell, R.L., et al.: Population growth and collapse in a multiagent model of the Kayenta Anasazi in Long House Valley. Proc. Nat. Acad. Sci. U.S.A. **99**, 7275–7279 (2002)
5. Sakahira, F., Terano, T.: Revisiting the dynamics between two ancient Japanese descent groups. In: Barceló, Juan A., Del Castillo, F. (eds.) Simulating Prehistoric and Ancient Worlds. CSS, pp. 281–310. Springer, Cham (2016). https://doi.org/10.1007/978-3-319-31481-5_10
6. Fujio, S.: The formation of Yayoi culture in Fukuoka plain: the interaction between hunter-gatherers and agrarian. Bull. Natl. Mus. Japan. Hist. **77**, 51–84 (1999). (in Japanese with English abstract)
7. Kataoka, K., Iizuka, M.: A mathematical study of the increase of the "Yayoi migrant population". Kyushu Koukogaku **81**, 1–20 (2006). (in Japanese with English abstract)
8. Nakahashi, T., Iizuka, M.: Anthropological study of the transition from the Jomon to the Yayoi periods in the Northern Kyushu using morphological and paleodemographical features. Anthropol. Sci. Jpn. Ser. **106**, 31–53 (1998). (in Japanese with English abstract)
9. Nakahashi, T., Iizuka, M.: Anthropological study of the transition from the Jomon to the Yayoi periods in the Northern Kyushu using morphological and paleodemographical features (2). Anthropol. Sci. Jpn. Ser. **116**, 131–143 (2008). (in Japanese with English abstract)

10. Epstein, J.M., Axtell, R.: Growing Artificial Societies: Social Science from the Bottom Up. Brookings Institution Press, Cambridge (1996)
11. Gilbert, N., Troitzsch, K.: Simulation for the Social Scientist. McGraw-Hill Education, New York (2005)
12. Bonabeau, E.: Agent-based modeling: Methods and techniques for simulating human systems. Proc. Nat. Acad. Sci. **99**, 7280–7287 (2002)
13. Tesfatsion, L.: Agent-based computational economics: growing economies from the bottom up. Artif. Life **8**, 55–82 (2002)
14. Tanaka, Y., Kikuchi, T., Kunigami, M., Yamada, T., Takahashi, H., Terano, T.: Classification of simulation results using log clusters in agent simulation. In: Second International Workshop of Artificial Intelligence of and for Business, JSAI International Symposium on AI 2017, p. 10 (2017)

Beyond Educational Policy Making

With Agent-Based Simulation

Atsushi Yoshikawa[1]([⊠]) and Satoshi Takahashi[2]

[1] Tokyo Institute of Technology, Yokohama-shi, Kanagawa 226-8503, Japan
`at_sushi_bar@dis.titech.ac.jp`
[2] Tokyo University of Science, Chiyoda-ku, Tokyo 102-0071, Japan

Abstract. In recent years, formulation of educational policy has come to be based on data. That data, however, can turn out to be difficult to access, or mixed with so much noise interfering with education policy formulation, that it cannot be used directly for policy making. To address this issue, an increasing number of attempts to contribute to policy formulation have been made using agent-based simulation (ABS). In the majority of research, ABS is used in the ex post facto analysis of why educational policy has not been effective. In this paper, case studies show that by incorporating ABS into the policy formulation process, the risk of failure can be reduced. By illustrating the relationships between model level, stage of educational policy formulation and the output scenarios of ABS, it is possible to determine which types of risks can be reduced. This paper presents ABS description levels, and discusses risks that both can and cannot be expressed using ABS. We show two ways to use ABS for educational policy making by identifying risks that can be reduced and risks that cannot be dealt with by ABS.

Keywords: Educational policy making · Agent-based simulation · Real-world data · Modeling of educational policy making

1 Introduction

Education policy not only affects the lives of individuals in the future. It also has a major effect on society, where the human resources produced by that education system are employed. The importance of education policy is widely acknowledged, and there are numerous reports describing outcomes of education policy in various countries, both successful and less effective. However, there are few studies on educational policy that explain how the policy has been derived and why it was considered valid at the time it was conceived. An educational policy that was a Japanese version of the No Child Left Behind (NCLB) Act was implemented in Japan about 20 years ago. Called 'yutori' education, or 'education with breathing space', the results of this policy have been assessed as contributing to the lowering of academic ability in Japan [1]. Some predicted the failure of this policy from the outset [2]. Even though there has been some self-reflection regarding the failure of the policy, to date there has been no research into the reasons why such a policy was adopted. Educational policies are determined after discussion in meetings and conferences, in the course of which data from different

© Springer Nature Singapore Pte Ltd. 2019
F. Koch et al. (Eds.): GEAR 2018, CCIS 999, pp. 112–130, 2019.
https://doi.org/10.1007/978-981-13-6936-0_12

studies are presented. If a policy leads to unanticipated results, an explanation should be sought. Educational policies that are less successful than envisaged are not limited to Japan. Educational policies related to the NCLB Act and the Perry Preschool Study were based on data and underwent examination but did not bring complete success [3, 4].

For this paper, we categorize precedents in agent-based simulation (ABS) related to educational policy as a method for formulating policy that offers a greater reduction in the risks of failure [5]. From that, we identify distinctive characteristics of ABS that can be linked with educational policy, and propose that introducing ABS into the process of formulating educational policy is likely to be effective.

2 Categorization of Educational Policy

2.1 Levels of Educational Policy

UNESCO states that the determination of educational policy (in the broad sense) involves three stages of (1) policy (referred to here in the narrow sense; hereafter, policy in this narrow sense will be referred to as 'policy measures' and in the broad sense it will be referred to as educational policy); (2) strategy, and (3) plans [6]. According to UNESCO's definition (here slightly revised), these are:

- A *policy* establishes the main goals and priorities pursued by the government in matters of education – at the sector and sub-sector levels – with regard to specific aspects such as access, quality and teachers, or to a given issue or need.
- A *strategy* specifies how the policy goals are to be achieved.
- A *plan* defines the targets, activities to be implemented and the timeline, responsibilities and resources needed to realize the policy and strategy.

Policy determines goals and priorities, typically determined by the external environment. In line with that determination, strategy devises methods for proceeding toward realization. Plans for concrete realization of those methods are, in short, the instantiation of policy in the form of targets and activities intended to achieve the strategy. UNESCO recommends that these three stages be carefully separated according to their function. Naturally, good policy measures do not appear full-blown from the very start, so it is envisaged that the four-part policy cycle of analysis–planning–implementation–evaluation will undergo numerous iterations. A look at this cycle will show, however, that though it is effective for improving an educational policy that has been decided on, it offers no guidance for determining new educational policy from the ground up. UNESCO itself further finds that effective educational policy has to be [6]:

- *Built on evidence*
- *Politically feasible*
- *Financially realistic*
- *Agreed to by the government and relevant stakeholders.*

While imposing these strict constraints, they state that the policy cycle is preceded by this vision:

Step 0 (Vision): Before the start of a policy cycle, a strategic intent, often called a "vision", is formed. For instance, once a political party wins a majority of seats in parliament and forms a government, they denote their strategic intent for education, which, for instance, may be: "Increase participation of youth from lower socio-economic backgrounds in tertiary education."

This is a weak point in the process of policy formulation.

2.2   Areas Addressed by Educational Policy

UNESCO delineates the following three levels of macro, meso, and micro as the areas addressed by educational policy [7].

(a)   Macro: region/state/national/international
(b)   Meso: institution-wide
(c)   Micro: individual user actions.

The evidence required to formulate policy differs depending on the area addressed, but this does not mean that the necessary evidence for the various areas exists. For instance, Japan's 'education with breathing space' system was formed by policy measures of the national government, and thus consists of macro policy measures. The system is set up so that matters are then decided in various derivative areas of policy with measures branching out from each level. These measures range from national curriculum standards to textbooks. Teacher training is the responsibility of the boards of education at the local government level [8]. A specific educational method was even devised for the classroom context, known as the integrated study period. This means that the process has been determined by policy and strategy. Following Japanese tradition, the plans are left up to the schools [9]. The shift toward 'education with breathing space' was probably carefully examined because there were data from subjective assessments conducted in the course of students' studies for entrance exam competitions, and qualitative data in the form of observations voiced by people involved in classroom education. However, no data exist on the time taken in teacher training. If data were available, it would be possible that some teachers experience unanticipated difficulties in adequately applying the approach to education [10]. These data would be part of planning, and may be difficult to obtain or does not exist. The result is that the information needed to enact educational policy is incomplete, and this is conceivably the reason that conditions emerged that were unanticipated by experts. This illustrates that when data supporting the level of the educational policy do not exist, unanticipated conditions tend to occur.

2.3   Data for Making Decisions About Educational Policy

Unless data for each of the policy–strategy–plan levels exist, the formulation of educational policy will be insufficiently grounded. However, it may also be unrealistic to obtain the full range of these data.

To formulate educational policy, real-world data from factual investigations of actual circumstances and other such sources are used. However, these real-world data may not be as useful as expected in informing educational policy, since such data may contain distortions due to various circumstances. For instance, there may be data showing that 100% of schools are equipped with personal computers. However, even if this is the case, there may be schools where computers are locked away in storage so that students cannot damage the expensive equipment, or where the computers are so old that they do not function properly and are of little use. In these cases, the computer adoption rate will be high, and if policy is made in the expectation that courses will involve the use of computers, then that policy will be out of touch with reality.

The real-world data used in formulating educational policy have the following limitations.

(a) Ethical considerations prevent control experiments from being performed, so the data do not contribute directly to policy aims.
(b) Real-world data are difficult to interpret because phenomena are overly complex.
(c) Long-term influences exist in data.
(d) Data cannot be acquired in the first place.

These are matters to be aware of at the stage of creating an educational policy. The weakness, however, is that measures to resolve the problems are difficult to come by. Item (d) is a problem that may not be possible to address, but the first three items are worth considering further.

Ethical Considerations Prevent Control Experiments from Being Performed, so the Data Do Not Contribute Directly to Policy Aims. When implementing policy measures, the scientific approach is to conduct experiments with the participants divided into control and experimental groups. Comparison of results from the two groups confirms the validity of the proposed method. These are known as control experiments. The field of education, however, faces the issue of equality of educational opportunity. This means that using control groups is not possible. At the same time, implementing an educational policy that cannot be expected to be effective is not ethically acceptable. Consequently, policy measures cannot be confirmed experimentally, even in part.

Real-World Data Are Difficult to Interpret Because Phenomena Are Overly Complex. The effectiveness of education is influenced not only by teachers and instructional materials, but also by the characteristics of the group of learners who receive the same education and by the environment. These factors exist across a range from the individual learner level to the macro level. At the individual learner level, factors may include the age and gender of the learner, the family structure, and so on. At the middle level, they may include the make-up of classes at the school or other educational institution, and the curriculum design. At the macro level, they may include the configuration of the school district, the placement of the educational institutions, and the system for advancing students to the next level of education. As this illustrates, the factors are interrelated at multiple levels in such complex ways that it is difficult to interpret acquired data.

Long-Term Influences Exist in Data. Developing human resources takes a long time. For that reason, there will not be just one educational policy that exerts an influence over that development period, and it is difficult to assess a policy and associated policy measures as a stand-alone unit. For instance, grades in lower secondary school mathematics are sometimes affected by insufficient acquisition of Japanese language competence in primary school [11]. Furthermore, a single educational policy can affect not only short-term changes in academic ability, but also long-term outcomes. There is the extremely rare case of the study of long-term influence in the Perry Preschool Study. This study of an early childhood education program in the United States, conducted from 1962 to 1967, carried out a continuous analysis of the influence of the educational program. The long-term influence continues to be studied even today [4]. Cases like this, however, are extremely unusual.

There is progress in addressing the deficiencies of data used for educational policy. With regard to the issue of ethical considerations, measures have been taken, for example, to use alternative experimental or quasi-experimental designs (e.g. crossover trials) and establish special districts related to education, e.g. charter schools in the United States [12]. An example of a quasi-experiment includes the assessment of educational subjects related to Title I schools, which makes use of differences between schools in their circumstances [13]. These are measures devised in the actual context of education in practice, and in that sense can be considered measures devised out of awareness of the limits involved.

2.4   Weaknesses in Educational Policy

A summary of results indicates that the course of formulating educational policy is subject to the following three weaknesses:

(a) When formulating new educational policy, we cannot expect to find real-world data that would serve as a reference.
(b) When policy-making lacks data that support the level of educational policy concerned, unanticipated conditions are likely to occur.
(c) When formulating educational policy, the real-world data that are used will have certain limitations.

3   Application of ABS to Educational Policy

3.1   Potential of ABS as Data

Gilbert states that the three weaknesses of data presented for the purpose of framing educational policy (referred to in Sect. 2.3), could be resolved by the use of ABS. This can be inferred from the distinctive characteristics of ABS listed below [5].

(a) As a simulation, it can be used to try out any configuration.
(b) It is specialized in the handling of complex systems.
(c) As a simulation, it allows the time interval to be freely selected.

As a Simulation, ABS Can Be Used to Try Out Any Configuration. In ABS, a virtual world is built and the agent is made to act within that world to carry out the simulation. The designers can try out any configuration they like in the virtual world. This distinctive characteristic makes it possible to conduct control experiments that would be difficult in the real world.

ABS is Specialized in the Handling of Complex Systems. With this technique, people are modeled as individuals, called agents, who act autonomously, and the real world in which people act is modeled as a virtual world. The simulation is then carried out by having the agents interact with each other and with the virtual world. This approach is therefore suited to dealing with complex phenomena caused by the interaction of elements at education-related levels ranging from the micro to the macro.

As a Simulation, ABS Allows the Time Interval to Be Freely Selected. In other words, the length of time taken to carry out the simulation can be configured freely as the designers wish. For that reason, this approach allows examination of the long-term influence of educational policy by means of simulation.

Due to these distinctive characteristics of ABS, data can be provided for the purpose of formulating educational policy that could not be achieved with real-world data alone. ABS is model-based, so it is created on the basis of abstract models in education. In other words, it is created with a variety of educational theories as its basis. Its parameters are derived from real-world data. The ABS results themselves are not used extensively. Rather, the various scenarios generated by ABS are interpreted to allow examination of: (a) the process by which the results of educational policy come about, (b) impact from the educational policy that is formulated, (c) divergence from expected values, and (d) how policy created to improve the situation might actually end up making it worse instead.

3.2   Model Levels

Gilbert explains that, like educational policy, ABS has three levels [5]. These are the abstract, middle range, and facsimile levels.

An abstract model has the lowest level of resolution. The model is constructed out of a small number of parameters and interactions, with the aim of gaining a fundamental understanding of a social phenomenon. A classic example is Schelling's segregation model [14].

The middle range model has a slightly higher resolution than the abstract model. The component elements and parameters are markedly more numerous than in the abstract model. One example that can be cited is the research of Yano et al. [15].

The facsimile model is the model with the highest resolution. It aims to recreate a social system under specific circumstances. One example is the model of a specific classroom.

According to the conceptual approach of this categorization, real-world educational policy consists of policy, strategy, and plan stages. They correspond to the facsimile model since they are created with awareness of specific circumstances. The ABS, by contrast, contributes to the formulation of educational policy by the interpretation of scenarios. Therefore it is a model of at least the middle range or higher (Fig. 1).

Figure 1 shows a schematic representation of the relationship between educational policy and ABS. It shows that ABS, through interpretation of scenarios, can yield data at the plan level of educational policy.

4 Cases of ABS Application to Educational Policy

4.1 Introduction of the Case Examples

Educational policy is formulated in a variety of fields. Here, therefore, it will be placed under the three classifications of macro, meso, and micro named in Sect. 2.2, and real-world educational policy and ABS will be viewed on that basis.

Model at Macro Level. Macro examples that could be cited include the NCLB Act, school system model, school district scope, and guidelines for establishing schools. The case of the NCLB Act is introduced here.

Case Example: About the No Child Left Behind Act. The NCLB Act was an educational policy in the United States that was passed into law under the administration of the country's 43rd president, George Walker Bush, in 2002 [16]. NCLB calls for student learning progress goals (adequate yearly progress: AYP) to be set state by state, and reading, math, and science are the areas for assessment. Progress toward goals is assessed by tests. Schools that do not achieve their goals are negatively viewed and required to take corrective action. The NCLB Act requires separate reports to be submitted for students of different income, race, disabilities, and English language learners. The purpose is to correct disparities in academic ability resulting from these factors, a distinctive characteristic of NCLB.

The US Department of Education has reported that various study results show links to improvements in students' academic ability [17]. For instance, the study results from the National Assessment of Educational Progress (NAEP) (https://nces.ed.gov/nationsreportcard/) are reported as follows:

> *For America's nine-year-olds in reading, more progress was made in five years than in the previous 28 combined.*
> *America's nine-year-olds posted the best scores in reading (since 1971) and math (since 1973) in the history of the report. America's 13-year-olds earned the highest math scores the test ever recorded.*
> *Reading and math scores for African American and Hispanic nine-year-olds reached an all-time high.*
> *Math scores for African American and Hispanic 13-year-olds reached an all-time high.*
> *Achievement gaps in reading and math between white and African American nine-year-olds and between white and Hispanic nine-year-olds are at an all-time low.*

However, various abuses have occurred, and in 2012, President Barack Obama, the 44th US president, decided to grant waivers [17]. 'Abuses' refer to the occurrence of conditions that had not been envisaged at the start of the plan. For instance, schools that were designated as 'failing' saw a decline in the morale of school staff and resignation

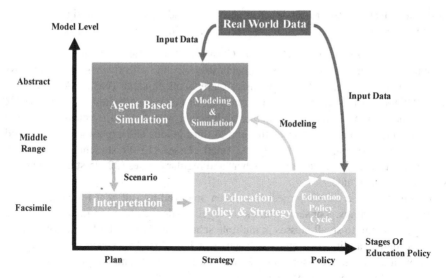

Fig. 1. Relationship between educational policy and agent-based simulation

rates increased [18]. In addition, since AYP assessment covers reading, math, and science, there was a tendency for other subjects not assessed to be neglected [19]. At the extreme, there were cases in which teachers of subjects other than those tested were fired and replaced by part-time instructors [20]. Schools that did not meet the pre-scribed levels were designated 'failing' schools. At schools that were designated as failing for two successive years, students were allowed to transfer to other schools – and conditions calling for transfer did actually occur.

Case Example of ABS Application: Causal Analysis of the NCLB Act by ABS. Sklar et al. suggest SimED as an example of ABS for use in examining educational policy [21]. In the SimED world, there are agents in the four roles of student, teacher, principal, and superintendent. These move around in a virtual world made up of the three levels of classroom, schoolhouse, and school district. The students have such variables as ability, motivation, emotion, belief set, satisfaction, family income, mobility, and performance. Using ABS with this kind of middle-range framework, Sklar et al. analyzed scenarios that could occur over a 20-year period following introduction of the NCLB. The introduction of a system of small class sizes (a 'STAR-like scenario' [22]) indicated the possibility that the percentage of students transferring to other schools could be lowered. This is primarily an ex post factor search for causes after a problem has occurred (in this case, scenario analysis). At the same time, however, Sklar et al. found that smaller class sizes would be a solution for the NCLB (Fig. 2).

Model at Meso Level. Meso level examples include curriculum design and class placement methods. The case of class size design will be introduced here.

Case Example: About Class Size Design. Class size design has been a matter of debate for many years, and large-scale experiments to verify the effectiveness of small class sizes have been carried out. For example, there is the Student Teacher Achievement Ratio (STAR) project in the state of Tennessee in the United States [19]. The STAR project tracked the changes in academic ability that were caused by class size in the four-year period from kindergarten through to third grade. The experiment was carried out with the following three class types: small classes (13–17 students per teacher), regular classes (22–25 students per teacher) and regular classes (22–25 students) with a full-time teacher's aide.

Results from the STAR project have been analyzed by different researchers, with contradictory findings. Konstantopoulos surveyed academic ability in mathematics and reading in relation to small classes at multiple schools, and found both cases of schools where academic ability rose and cases where it fell [23]. The situation with regular classes with a full-time teacher's aide was similar: the results showing instances where academic ability rose and other examples of where it fell.

There was also a five-year pilot project in the state of Wisconsin that started in the 1996/1997 academic year. Known as the Student Achievement Guarantee in Education (SAGE) project, this involved intervention experiments with measures such as lowering the number of students per teacher at participating schools to 15 [24]. The data collected were analyzed using regression analysis and hierarchical linear models (HLM), producing the conclusion that a small class size is effective in raising students' academic ability. The approach was particularly effective for students of African-American descent, indicating the possibility that having small class sizes could become an effective means of offsetting the difference in academic ability between African-American and white students.

Based on research to date, a conclusion on whether small class sizes are effective or not has yet to be reached.

Case Example of ABS Application: Analysis of Class Size Design by ABS. Yano et al. proposed an ABS model to use in analyzing class size [15]. The model contains agents in the two roles of student and teacher. These agents move and act in the same world, but the students and teachers belong to different schools, either primary, lower secondary, or upper secondary school. Multiple schools exist, and when students advance to the next level, they go to nearby schools. Students are characterized by variables such as academic ability, learning strategies, and academic motivation. Learning strategies follow three styles: teacher-driven, mutual teaching, and motivation-dependent. The students therefore proceed with their learning activities in accordance with their learning strategies, teaching each other, being educated by their teacher, and so on.

Yano et al. examined the scenarios that would be generated by where teacher assignments were concentrated in primary school, lower secondary school, or upper secondary school, at which small class sizes were realized. The results indicated that improvements in academic ability could be obtained among students at the lower levels of academic ability using the teacher-driven and mutual teaching styles of learning

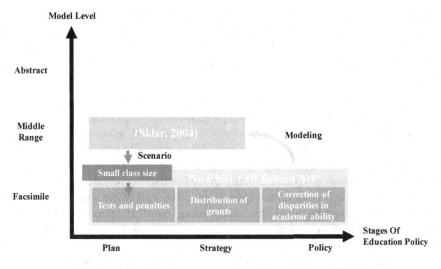

Fig. 2. Positioning the answers yielded by NCLB and SimED

when additional teacher agents were assigned to the primary schools. With the motivation-dependent style, however, these results were not observed. This indicates that results will differ according to student learning strategies. These results concur with the research of Konstantopoulos and Molnar [23, 24].

Interpreting these results, we find that having a system of small class sizes is given in educational policy, and that strategies and plans are positioned as practices at the educational site. In ABS, on the other hand, the model is positioned to place emphasis on the plan. This relationship is shown in Fig. 3.

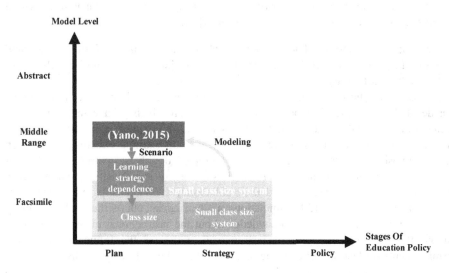

Fig. 3. Small class size in relation to solutions derived by the model of Yano et al.

Model at Micro Level. Micro level examples include design methods for collaborative and cooperative learning. Strict distinctions between collaborative and cooperative learning have been pointed out. Here, however, these are positioned as attempts to heighten learning effectiveness by having students influence each other through group learning [25].

Case Example: Collaborative Learning. At public schools in the United States during the 1970s, measures to eliminate racial segregation were implemented. The mistrust between whites and African-Americans was deeply rooted, however, and frequent clashes occurred. Aronson proposed the jigsaw method to build relationships of mutual trust during the primary education phase, when mistrust is relatively small [26]. In the jigsaw method, students are placed in small groups of five to six and members of each group are further placed in separate expert groups. The goal of the small groups is for each one to solve a problem together. The students are first separately given information in their respective expert groups. They are called on to assimilate that information and carry it back to their small groups. Next, the students are called on to integrate their pieces of information and solve their assigned problems. The problems are designed to yield solutions when the various pieces of information are well integrated. The aim is to have cooperative relationships form naturally among the students.

Aronson finds that use of the Jigsaw method leads to the formation of cooperative relationships as a result of learning: (1) that within their groups, nobody is able to learn without the help of others; and (2) that the members of the groups are able to make contributions that are uniquely their own and are essential.

Analysis of Collaborative and Cooperative Learning by ABS. Here we introduce the ABS approaches of Kuniyoshi et al. and Spoelstra et al. Kuniyoshi et al. examined the effectiveness of learning in terms of learners' academic ability, the structure of their instructional materials, and their collaborative relationships [27]. Spoelstra et al. examined the impact on group learning results from diversity of capabilities within groups, group sizes, and rewards to the groups [28].

Case example 1: Model of knowledge teaching on a complex doubly structural model taking academic ability, instructional material structure, and collaborative relationships into account

Kuniyoshi et al. did not use the jigsaw method in their research, but have modeled the inside of the Japanese classroom. Through this approach they have investigated the extent of influence that learners' individual academic ability, their instructional material structure, and their collaborative relationships have on learning effectiveness [27]. In the Kuniyoshi et al. model, agents use a 'complex doubly structural network' that models the structure of the learners' understanding of their instructional materials (knowledge network structure) and their collaborative relationships with each other (social network structure) [29]. Simulations are then carried out on the academic ability of the learners, the structure of knowledge about the instructional materials that they study, the teaching strategies of instructional staff, and the influence of seating arrangements on learning effectiveness. This yielded the following results:

(a) *When different teaching strategies, seating arrangements, and collaborative learning are used, learning effects vary.*

(b) *Group-style collaborative learning with a dispersed seating arrangement has the best effect on learning, and this would be followed by the method with the highest answer rate.*

(c) *An ability-based class has a negative effect on collaborative learning because diversity is reduced, and learner homogeneity decreases the collaborative effect.*

(d) *If teaching occurs only once for one knowledge item, learners may fall behind. Reviews should be conducted more frequently to facilitate the anchoring of the knowledge in a class.*

Case example 2: Diversity of group composition is problem-solving capability

Spoelstra et al. conducted analysis of learning effects in Student Teams Achievement Divisions (STAD) using ABS [28, 30]. The flow of learning in STAD is as follows [28]:

(a) *Teacher presentations: The initial phase of the learning process in which a teacher explains the concept to be acquired;*

(b) *Student teamwork or individual work: The phase in which activities designed to facilitate learning are undertaken by one or more students, working alone or in groups;*

(c) *Quizzes: The phase in which the teacher evaluates the progress made by each student;*

(d) *Individual improvement: The phase in which individuals receive recognition (from the teacher and/or their peers) for any progress they have made; and, optionally,*

(e) *Team recognition: The phase in which teams are ranked and prizes (or some other form of recognition) are bestowed upon team members. This phase is only relevant when the 'cooperative goal structure' is in place and students are working in teams.*

The models used by Spoelstra et al. have student agents. Student agents have state variables such as ability, emotion, motivation, and zone. They engage in learning in accordance with the STAD model and obtain rewards commensurate with their academic performance.

Spoelstra et al. used this ABS model to analyze diversity in the levels of capability in the group, group size, and rewards to the group, and how these influence learning effectiveness in STAD. This analysis indicated that diversity, group size, and rewards to the group exert influences on group learning effectiveness. In a particularly startling result, Spoelstra et al. emphasized the possibility that reward might be effective for a group made up entirely of people with low levels of ability and if the reward was of an appropriate size.

The relationship between educational policy and the Kuniyoshi ABS is shown in Fig. 4. There is less of a correspondence here than was shown in the macro and meso cases. However, what Kuniyoshi refers to as the comprehension of instructional materials is something the jigsaw expert group learned well, and if collaborative relationships are thought of as small jigsaw groups, then Kuniyoshi's ABS data could be interpreted as identifying further success factors contributing to the effectiveness of the jigsaw approach (Fig. 5).

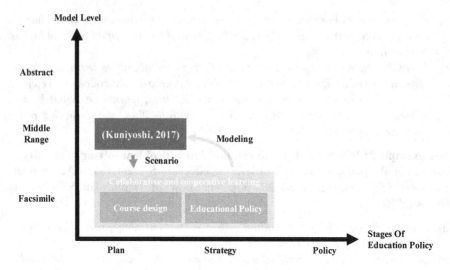

Fig. 4. Relationship between collaborative learning and the explanation of the Kuniyoshi et al. model.

4.2   Consistency of Educational Policy with ABS

The simulation results of ABS, representing the actuality of educational policy, can explain the results of educational policy implementation. However, the interpretation of ABS results may intrude because the model providing the basis for ABS is at a higher level of abstraction. This is because the data used for educational policy and educational structure provide the source for ABS. The data that are necessary in the formulation of educational policy can be configured as parameters in ABS, and the scenario will change depending on that configuration. Consequently, the particular scenario that is adopted will change the results that are obtained. Furthermore, even for just a single scenario, the simulation can be carried out for an enormous number of iterations so that it can find unexpected risks. This approach is able to perceive the pitfalls that lie in the path of educational policy. Arai et al. used ABS to derive the decline in academic ability that had not been envisaged when 'education with breathing space' was being designed [31].

Whether the target of the educational policy is at the macro, meso, or micro level, the positioning with regard to educational policy and ABS is maintained in the relationship shown in Fig. 1. When it comes to the interpretation of ABS results, however, depending on the ABS model, it may: (a) relate to specific policy measures at the plan level, or (b) turn out to be capable of explaining the reason why an educational policy brought about the results it did. The case examples show that ABS can be used in these two orientations.

5 Change in the Course of Educational Policy Formulation Due to ABS

In Sect. 2, the following three areas were identified as weaknesses occurring in the course of formulating educational policy:

(a) When formulating new educational policy, we cannot expect to find real-world data that would serve as a reference.
(b) When policy-making lacks data that support the level of educational policy concerned, unanticipated conditions are likely to occur.
(c) When formulating educational policy, the real-world data that are used will have certain limitations.

In Sects. 3 and 4, the potential of ABS to make up for these weaknesses was discussed. Based on the case examples, ABS was found to have two main capabilities:

(a) To explain the reason why an educational policy brought about the results it did;
(b) To examine specific policy measures at the plan level.

These findings will be placed into organized form as a method for using ABS in the course of formulating educational policy.

5.1 Application of ABS in the Course of Formulating Educational Policy

Examination of the process of educational policy formulation when that process is underway will produce something similar to Fig. 5. With regard to the first capability, the formulation of educational policy means that the ABS model can be structured on that basis. The results of that simulation can then be crosschecked with the data generated by implementation of the educational policy. Interpreting both sets of

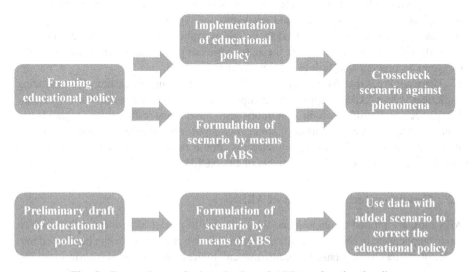

Fig. 5. Two pathways for introduction of ABS to educational policy

findings, makes it possible to explain the results. With regard to the second capability, the process of framing policy can be formulated as a scenario in ABS based on the preliminary draft of the policy. The scenarios generated by ABS are checked for unexpected risks. If necessary, the formulation of the scenario can also be repeated in ABS and corrections made.

In this way, methods for using ABS can be envisaged as Usage Method (1): Causal Analysis by ABS, and Usage Method (2): Identifying Scenarios by ABS. By employing ABS, causal analysis of educational policy can be carried out rapidly even after the policy has been formulated. In addition, formulating a scenario at the preliminary draft stage of a policy, can enable correction of the preliminary draft of a policy by incorporating matters that policy-makers could not have foreseen. Scenarios can be examined to determine whether they fall within what was envisaged, and are within the acceptable limits of risk.

Specifically, Usage Method (1) applies ABS to causal analysis of the reasons why education is in its current state, and of the mechanism by which an implemented educational policy led to its effect, or did not lead to an effect. For instance, verification of the effectiveness of a small class size also indicated the possibility that the variations in academic ability within a class are related to improvement in the academic ability of the class as a whole [32]. In this case, it was reasoned that exchange between students with higher academic ability and those with lesser abilities in a small class heightened academic motivation on both sides. By applying ABS in this situation, it is possible to analyze in greater detail situations such as how students with different levels of academic ability help each other.

In Usage Method (2), ABS is used for the examination of possible ways in which the preliminary draft of an educational policy can lead to results. By using ABS, scenarios that could be brought about by the educational policy can be identified, and on that basis, the educational policy improved and implemented. In connection with the NCLB Act, for instance, it is possible to analyze the types of scenarios that could occur in future as a result of state-by-state differences in fiscal resources, the economic status of residents, racial composition, and academic ability. Specifically, this examines the combination of conditions that lead to failure of a policy or a rise in student dropout rates. By identifying scenarios like these, scenarios of policy failure are discovered and measures to prevent that failure can be examined.

There are areas that ABS cannot handle such as rigorous forecasting and producing numerical values. ABS is applied to complex systems, and is sensitive to initial conditions. A small difference in conditions can yield great differences in results. In connection with the examination of small class sizes, for instance, this approach is not able to deciding exactly how many students should be in a class. It is also unable to forecast the specific quantitative effects of the NCLB Act.

5.2   Conditions for Application of ABS

ABS must be applied to phenomena that it is able to model. Considering the cases of model application in education analyzed by Yoshikawa et al., models can be divided into three categories: models based on mathematical, logical, and structural descriptions [33].

Models based on mathematical description are models that express the relationship between their component elements using numerical formulae. Models based on logical descriptions are models that express the logical relationship of their component elements. Logical relationships are expressed by figures and text that supplement these figures. Models based on structural descriptions refer to models that decompose their component elements and express the content of those elements as well as their structure and categorization. Structure and content are expressed by figures and text that supplement those figures (Fig. 6).

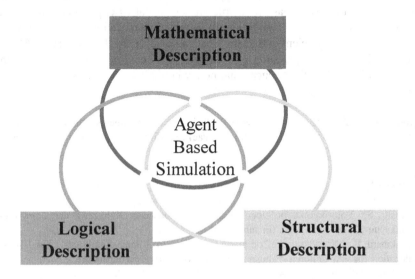

Fig. 6. Systematic ordering of models handled in education

An agent does not belong to one of the three categories. Basically, it has a logical or structural description (or both) and it is re-expressed with a mathematical description so that it can be incorporated into agent behavior. In other words, it is created from a combination of at least two or more descriptions. To apply ABS, the phenomenon must be such that at least a structural or logical description can be made and incorporated into a mathematical description. When Yano et al. modeled the influence between students and teachers, they created a logical description model of the relationship between various elements relating to student learning and the amount of learning. After that, they created a mathematical description model to express it.

As an example of educational policy implementation, on the other hand, results have been obtained indicating that small class sizes contribute to improvement of the students' self-image with respect to their academic learning [34]. Since self-image is difficult to describe by a mathematical model, however, it cannot be incorporated into the process of formulating educational policy using ABS.

6 Conclusion

This paper has discussed the use of ABS in addressing questions about why specific outcomes occurred following policy implementation, and whether a process of formulating education policy can be created to reduce the risks in formulation of policy. Answers were derived from cases in which the ABS approach was realized in ways that matched relatively closely with educational policy.

ABS was positioned as a middle-range model, and certain limitations were noted. In terms of the level of description, the approach is limited to policy addressing matters that are subject, at the very least, to a structural or logical description and that can be rewritten into a mathematical description. In formulating educational policy, however, the data are often insufficient and many challenges arise. In that context, ABS is a powerful tool for the interpretation of scenarios. If an environment was created that supports a tool like this, then educational policy could undergo major changes in the future.

Acknowledgments. We thank all reviewers for their valuable comments, and Michelle Pascoe, PhD, from Edanz Group (www.edanzediting.com/ac) for editing a draft of this manuscript.

References

1. Tadashi, S.: Yutorikyoiku no rinen ni kannsuru kousatstu. Kanagawa daigaku·kyouiku kennksyuu **32**, 49–53 (2012). (in Japanese)
2. Sakakibara, E.: "Yotorikyouiku" de nihon metsubou. In: Bungeishunju (eds.) The Issues in Education for you Children. Bungeishunju, Tokyo (2001). (in Japanese)
3. PUBLIC LAW 107–110—JAN. 8, 2002 (2002)
4. Belfield, C.R., Nores, M., Barnett, S., Schweinhart, L.: The high/scope perry preschool program cost–benefit analysis using data from the age-40 followup. J. Hum. Resour. **41**(1), 162–190 (2006)
5. Gilbert, N.: Agent-Based Model, Quantitative Applications in the Social Sciences, vol. 153. SAGE Publications, London (2007)
6. UNESCO Bangkok: UNESCO Handbook on Education Policy Analysis and Programming, Volume 1: Education Policy Analysis. UNESCO Bangkok (2013)
7. Shum, B.S.: UNESCO Policy Brief: Learning Analytics. Published by the UNESCO Instute for Information Technologies in Education, Moscow (2012)
8. Textbook Publishers Association of Japan, Gakusyusidouyouryoutozissizikitokyoukasy-onohakkannzikitinokannkeinituitesetumeisitekudasai.   http://www.textbook.or.jp/question/answer/a04.pdf. Accessed 20 June 2018. (in Japanese)
9. Ministry of Education, Culture, Sports, Science and Technology, Syougakkou gakusyuu sidou youryo, December 1998. http://www.mext.go.jp/a_menu/shotou/cs/1319941.htm. Accessed 20 June 2018. (in Japanese)
10. Takayama, K.: A nation at risk crosses the Pacific: transnational borrowing of the US crisis discourse in the debate on education reform in Japan. Comp. Educ. Rev. **51**(4), 423–446 (2007)
11. Ministry of Education, Culture, Sports, Science and Technology, Gakusyusidoyouey-outounokyouikukateinokizyuntounoarikatanitsuite,   http://www.mext.go.jp/b_menu/shingi/chukyo/chukyo3/021/siryo/attach/1399083.htm. Accessed 20 June 2018. (in Japanese)

12. Hanushek, E.A., Kain, J.F., Rivkin, S.G., Branch, G.F.: Charter school quality and parental decision making with school choice. J. Public Econ. **91**(5), 823–848 (2007)
13. Saunders, W.M., Goldenberg, C.N., Gallimore, R.: Increasing achievement by focusing grade-level teams on improving classroom learning: a prospective, quasi-experimental study of Title I schools. Am. Educ. Res. J. **46**(4), 1006–1033 (2009)
14. Schelling, T.C.: Dynamic models of segregation. J. Math. Sociol. **1**(2), 143–186 (1971)
15. Yano, Y., Kanzawa, A., Yamada, T., Yoshikawa, A., Terano, T.: Does small-size class improve academic abilities of students? —An analysis by agent-based approach—. Trans. Jpn. Soc. Inf. Syst. Educ. **32**(4), 236–245 (2015). (in Japanese)
16. No Child Left Behind Act Is Working. https://www2.ed.gov/nclb/overview/importance/nclbworking.pdf. Accessed 20 June 2018
17. Riddle, W.: What Impact Will NCLB Waivers Have on the Consistency, Complexity and Transparency of State Accountability Systems?. Center on Education Policy, Washington, DC (2012)
18. Finnigan, K.S., Gross, B.: Do accountability policy sanctions influence teacher motivation? Lessons from Chicago's low-performing schools. Am. Educ. Res. J. **44**(3), 594–630 (2007)
19. Dillon, S.: Schools Cut Back Subjects to Push Reading and Math. New York Times, New York (2006)
20. Meier, D.: Many Children Left Behind: How the No Child Left Behind Act is Damaging Our Children and Our Schools. Beacon Press, Boston (2004)
21. Sklar, E., Davies, M., Co, M.S.T.: SimEd: simulating education as a multi agent system. In: Proceedings of the Third International Joint Conference on Autonomous Agents and Multiagent Systems-Volume 3, pp. 998–1005. IEEE Computer Society, Washington, DC (2004)
22. Word, E., et al.: Student/Teacher Achievement Ratio (STAR) Tennessee's K-3 Class Size Study. Final Summary Report 1985-1990. Tennessee State Department of Education (1990)
23. Konstantopoulos, S.: How consistent are class size effects? Eval. Rev. **35**(1), 71–92 (2011)
24. Molnar, A., Smith, P., Zahorik, J., Palmer, A., Halbach, A., Ehrle, K.: Evaluating the SAGE program: a pilot program in targeted pupil-teacher reduction in Wisconsin. Educ. Eval. Policy Anal. **21**(2), 165–177 (1999)
25. Johnson, R.T., Johnson, D.W.: Cooperative learning in the science classroom. Sci. Child. **24**(2), 31–32 (1986)
26. Aronson, E., Patnoe, S.: Cooperation in the Classroom: The Jigsaw Method, 3rd edn. Pinter & Martin Ltd., London (2011)
27. Kuniyoshi, K., Kurahashi, S.: How do children learn and teach? In-class collaborative teaching simulation on the complex doubly structural network. SICE J. Control. Meas. Syst. Integr. **10**(6), 520–527 (2017)
28. Spoelstra, M., Sklar, E.: Agent-based simulation of group learning. In: Antunes, L., Paolucci, M., Norling, E. (eds.) MABS 2007. LNCS (LNAI), vol. 5003, pp. 69–83. Springer, Heidelberg (2008). https://doi.org/10.1007/978-3-540-70916-9_6
29. Kunigami, M., Kobayashi, M., Yamadera, S., Yamada, T., Terano, T.: A doubly structural network model and analysis on the emergence of money. In: Takadama, K., Cioffi-Revilla, C., Deffuant, G. (eds.) Simulating Interacting Agents and Social Phenomena, Agent-Based Social Systems, vol. 7, pp. 137–149. Springer, Heidelberg (2010). https://doi.org/10.1007/978-4-431-99781-8_10
30. Slavin, R.E.: Student teams and achievement divisions. J. Res. Dev. Educ. **12**(1), 39–49 (1978)
31. Arai, A., Terano, T.: *Yutori* is considered harmful: agent-based analysis for education policy in Japan. In: Shiratori, R., Arai, K., Kato, F. (eds.) Gaming, Simulations, and Society, pp. 129–136. Springer, Tokyo (2005). https://doi.org/10.1007/4-431-26797-2_14

32. Mitchell, D.E., Beach, S.A., Badaral, G.: Modeling the relationship between achievement and class size: a re-analysis of the Tennessee project star data. Peabody J. Educ. **67**, 34–74 (1989)
33. Yoshikawa, A., Takahashi, S.: Review on educational application of agent technologies. Trans. Jpn. Soc. Inf. Syst. Educ. **35**(1), 5–12 (2018). (in Japanese)
34. Mulkey, L.M., Catsambis, S., Steelman, L.C., Crain, R.L.: The long-term effects of ability grouping in mathematics: a national investigation. Soc. Psychol. Educ. **8**(2), 137–177 (2005)

Author Index

Printed in the United States
By Bookmasters